Prunus

Edited by Ayzin Küden and Ali Küden

Published in London, United Kingdom

IntechOpen

Supporting open minds since 2005

Prunus
http://dx.doi.org/10.5772/intechopen.87486
Edited by Ayzin Küden and Ali Küden

Contributors
Madalina Butac, Dario Chavez, Jose Chaparro, Thouraya Azizi, Youssef Ammari, Naira Ashraf, Moieza Ashraf, Siddhesh Bhende, Esonye Chizoo, Onukwuli Okechukwu Dominic, Akuzuo Ofoefule, Ume Sunday Cyril, Ogbodo Jacintha Nkiruka

First published in London, United Kingdom, 2021 by IntechOpen
IntechOpen is the global imprint of INTECHOPEN LIMITED, registered in England and Wales, registration number: 11086078, 5 Princes Gate Court, London, SW7 2QJ, United Kingdom
Printed in Croatia

British Library Cataloguing-in-Publication Data
A catalogue record for this book is available from the British Library

Additional hard and PDF copies can be obtained from orders@intechopen.com

Prunus
Edited by Ayzin Küden and Ali Küden
p. cm.
Print ISBN 978-1-83962-607-4
Online ISBN 978-1-83962-608-1
eBook (PDF) ISBN 978-1-83962-609-8

For EU product safety concerns:
IN TECH d.o.o., Prolaz Marije Krucifikse Kozulić 3, 51000 Rijeka, Croatia, info@intechopen.com or visit our website at intechopen.com.

Meet the editors

Ayzin B. Küden is a full professor, physiologist, and stone fruit breeder, as well as being in charge of the peach breeding program and selection and collection of fig, apricot, and almond germplasm at the University of Cukurova. She has been working for 37 years on the physiological studies of temperate fruit trees under subtropical conditions, chilling requirements and dormancy breaking, resilience to climate change, peach breeding, and releasing new cultivars. She has published more than 190 national and international scientific papers and books. She was the convener of several national and international symposia, congresses and workshops and chairperson of TFTS Working Group in the International Society for Horticultural Science. She is currently working on 9 national and 2 international scientific projects.

Ali Küden is a full professor, specialized in temperate fruit growing, propagation, pruning, training, pomology and physiology at the University of Cukurova. He has been working for 39 years, especially on the propagation and training of mainly pome and stone fruits and also nuts both under continental and subtropical climatic conditions. He was the Director of the Pozanti Agricultural Research and Application Center of the University of Cukurova from 1997 to 2017. He has published more than 180 national and international scientific papers and books. He was the convener or co-convener of several national and international symposia, congress and workshops. He is currently working on 8 national and 1 international scientific projects.

Contents

Preface

The book "Prunus" contains chapters on breeding, germplasm, fruit tree physiology, and production of Prunus species, written by authors from different parts of the world. Prunus is one of the most important fruit genera widely spread according to the various climatic and soil conditions. This wide adaptability of the Prunus genus gives an opportunity for it to be grown in many parts of the world. In modern taxonomy, subgenera of Prunus such as Amygdalus, Cerasus, Laurocerasus, Lithocerasus, Padus and Prunus include many species among which Prunus persica L., Prunus domestica L., Prunus armeniaca L., Prunus avium L. are the main ones.

The mainland of peach (Prunus persica L. Batsch) is China; however, the production of peach has been extended to the different parts of the world in various ecologies for fresh and fruit juice consumption. It is grown at 25-45° north and south latitudes.

Plum (Prunus domestica L.) is one of the stone fruit species that has spread and adapted to different climatic regions of the world with a high number of species and cultivars. The reason for the extension of plums in such a large area is the suitability of plum to various conditions.

Sweet cherry (Prunus avium L.) originated in Southern Caucasus, the Caspian Sea and the North-Eastern Anatolia. These gene centers spread to the east and west covering a wide area of the world. Sweet cherry is abundantly found in the wild areas of North Anatolian and the Taurus Mountains in Turkey.

Apricot (Prunus armeniaca L.) is a stone fruit with its origin in Central Asia, Western China and Iran-Caucasus. It is economically cultivated in many countries in the world, especially in the Mediterranean countries. Adaptation of apricot to different climatic conditions is weaker than peach.

Briefly, this book is on the Prunus species which is one of the main fruit and nursery plants grown in the world.

Dr. Ayzin B. Kuden and Dr. Ali Kuden
Professor,
University of Cukurova,
Turkey

Section 1

Breeding and Genetic Resources of Prunus Species

Chapter 1

Plum Breeding

Madalina Butac

Abstract

Worldwide, plum is one of the main species, occupying an area of about 2,600,000 ha and ensuring production about 11,700,000 tons. Even though there are over 6000 plum cultivars belonging to 19–40 species, there is still the need to create new cultivars due to the demands of growers and consumers. In addition, the large plum-growing countries (Romania, Serbia, Germany, Bulgaria, etc.) have decreased production due to plum pox virus (PPV) attack. Therefore, these countries developed breeding programs with the following objectives: resistance/tolerance to PPV, productivity, fruit quality, late blooming, self-fertility, different ripening times, short growing period, spur fructification, etc. Using different breeding methods (controlled hybridization, open pollination, selection in wild population on *Prunus* sp., and mutagenesis), in the last years, over 450 plum cultivars were released, from which 70% represent European cultivars and 30% Japanese cultivars.

Keywords: plum, breeding, objectives, genetic resources, achievements

1. Introduction

Plum is one of the main groups of fruits with about 6000 cultivars, belonging to 19–40 species, originating from Europe, Asia, and the USA [1–5].

Plums are appreciated for fresh consumption and also for dehydration and processing in the different forms (jams, compotes, jellies, candied fruits, frozen fruits, liqueurs, brandy, etc.) [6].

Plums are the fruits with the highest nutritional value, having a high content in carbohydrates, minerals, and vitamins that stimulate the body's health [3, 7].

Fresh fruits contain sugar (16–20%), proteins (0.7%), lipids (0.28%), pectins and tanoid substances, etc. Dehydrated fruits have a high content in sugar in competition with figs and jujube (**Table 1**).

Recent studies at Tuft University, Boston, have shown that dehydrated plums have the highest antioxidants content, contributing to the neutralization of free radicals and thus to the prevention of cancer.

The therapeutic and prophylactic value of plums has been known since ancient times; they have alkalizing, mineralizing, laxative, diuretic actions.

Plum genetics and breeding have been reviewed over time by different specialists: Cullinan (1937), Weinberger (1975), Ramming and Cociu (1991), Okie and Weinberger [8], Okie and Ramming [9], Okie and Hancock (2008), Hartmann and Neumuller (2009), Neumuller (2010), Topp et al. [5], Butac et al. [10], Milosevic and Milosevic [3]. During the time, in the plum breeding programs, these researchers change different knowledge, breeding techniques, biological material, etc.

	Water content %	Pectic substances (%)	Fat substances (%)	Carbohydrates (%)	Calories					
Flesh fruits	87	0.7	0.28	16-20	75					
Dehydrated fruits	28	2.1	0.6	65-70	255					
		Vitamins (mg %)			Mineral substances (mg %)					
	A	B1	B2	B6	C	Ca	P	Fe	Na	K
Flesh fruits	350	0.03	0.03	0.5	10	12	18	0.5	0.1	170
Dehydrated fruits	1600	0.09	0.17	0.2	3.0	51	79	3.9	8.0	604

Table 1.
Chemical composition of plums.

2. Economic importance

Currently the world area planted with plum is 2,619,471 ha, which is up from the 1980s. Of the continents, the largest plum growing are Asia (2,125,006 ha) and Europe (353,919 ha) (**Figure 1**).

The situation in the countries is as follows: China 1,987,284 ha; Serbia, 72,024 ha; Romania, 66,680 ha; and the USA. 25,500 ha (**Table 2**).

In the world production, plum has about 2%, practically a modest place. As a temperate species, it still occupies the fourth place, after the apple, pear and peach, in this area. Plum production increased to 11,758,135 tons in 2017 (**Figure 2**). Due to the main contribution to this growth, Asia has practically become the largest producer of plums (8201 million tons, respectively, 69.75% of the world production), followed by Europe (2199 million tons, respectively, 18.70% of the world production), the USA (423 million tons), etc. (**Table 2**). It should be noted that while the average plum yield in Asia increased, the one in Europe has decreased by 5%.

According to data reported in December 2019 by FAO Yearbook, the largest producing countries (in thousand tons) are China (6804), Romania (434), the USA (423), Serbia (330), Turkey (292), Italy (207), France (205), Ukraine (200), Spain (172), the Russian Federation (146), the Republic of Korea (83), the Republic of Moldova (76), Bosnia and Herzegovina (74), Poland (58), Hungary (43), Bulgaria (49), and Germany (24) (**Table 2**).

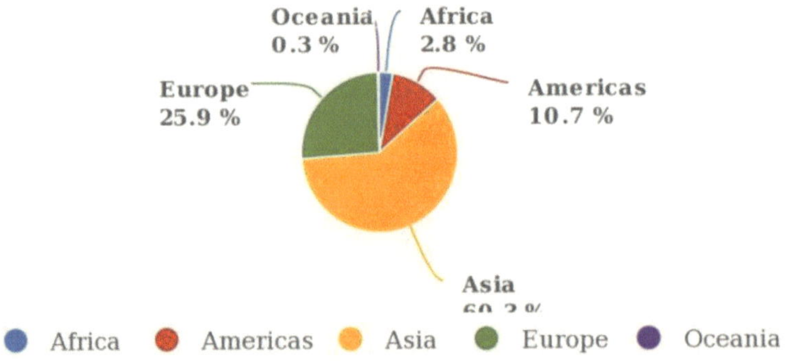

Figure 1.
Production share of plums and sloes by region: average 1994–2017. Source: [11].

World/country	Surface (ha)	Production (thousand tons)
Total	**2,619,471**	**11,758**
Asia	2,125,006	8201
Europe	353,919	2199
SUA	25,500	423
China	1,713,600	5664
Romania	66,680	434
Serbia	72,024	330
Turkey	21,385	292
France	16,958	205
Italy	12,676	207
Ukraine	18,000	200
Spain	15.199	172
Russian Federation	36,442	146
Republic of Korea	7495	83
Republic of Moldova	15,955	76
Bosnia Herzegovina	38,081	74
Poland	14,344	58
Bulgaria	6805	49
Hungary	7980	43
Germany	4191	24
United Kingdom	640	13

Table 2.
Area and production of plums in the world and in major growing countries.

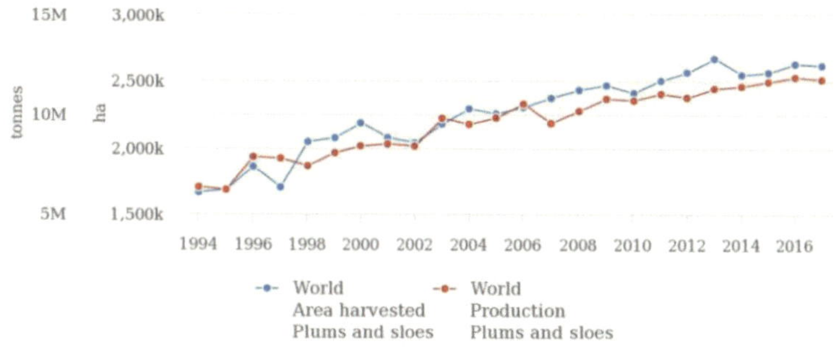

Figure 2.
Production/yield quantities of plums and sloes in the world (total): 1994–2017. Source: [11].

China is the country with the fastest development in the plum production, but the major producers of plums, from *Prunus domestica* and *Prunus insititia* species, are Romania, Serbia, the Republic of Moldova, Spain, Russia, Poland, France, Germany, and Bulgaria alarming declines in plums production. These decreases are due to the competition of the citrus, peaches, bananas, and other fruits but especially to the viral diseases that have destroyed the plum orchards and depreciated fruit quality in those countries.

3. Origin and history of plum

Plums belong to the genus *Prunus* L., subfamily Amygdaloideae (syn. Prunoideae), and family Rosaceae Jussieu [3, 12]. The basic chromosome number of plum is 8 (x = 8). Plum is a member of *Prunophora* subgenus, which itself is subdivided into the sections Prunocerasus and Euprunus. The Prunocerasus section contains the following species: *P. americana*, *P. angustifolia*, *P. hortulana*, *P. munsoniana*, *P. mexicana*, *P. nigra*, and *P. maritima*. The Euprunus section contains the following species: *P. domestica*, *P. spinosa*, *P. cerasifera*, *P. salicina*, *P. cocomilia*, *P. insititia*, *P. simonii*, and *P. ussuriensis* (**Table 3**).

Plum has a large spreading area. According to Vavilov's research, there are three spreading centers for plum species: Euro-Asian, North American, and East Asian [13].

In the Euro-Asian center, the following species are present: *Prunus domestica*, *Prunus insititia*, *Prunus spinosa*, and *Prunus cerasifera*, which are widespread in South Europe, Western Asia, around the Caucasus Mountains, and Caspian Sea, but also in the Balkans, as well as in the Mediterranean countries.

In the North American center which starts from the Gulf of Mexico and the West coast of the USA to Canada in the North, the following species are spreading: *Prunus nigra*, *Prunus americana*, and *Prunus munsoniana*.

The third center, East Asian, includes the following species: *Prunus ussuriensis*, *Prunus salicina*, and *Prunus simonii*.

From this large diversity, the most important species in commercial orchards are European or domestic plum (*Prunus domestica* L.; hexaploid species with 2n = 6x = 48) and Japanese plum (*Prunus salicina* Lindl.; diploid species with 2n = 4x = 16) [4, 5, 8, 14, 15]. European and Japanese plums belong to the same taxonomic section, but they are differentiated by origin and requirements to environmental factors.

The European plum is the most important plum in Europe, but it is also grown on other continents. The origin place of this species is Caucasus Mountains near the Caspian Sea [3]. These species grown in cooler areas can be divided into several groups considering the fruit characters: plumes, prunes, greengages or reineclaudes and mirabelles [16].

Crane and Lawrence [17] suggested that *Prunus domestica* (2n = 6x = 48, genome formula CCSSSS) is genetically a hybrid between diploid cherry plum (*Prunus cerasifera* Ehrh. 2n = 2x = 16, genome formula CC) and tetraploid sloe or blackthorn (*Prunus spinosa* L., 2n = 4x = 32, genome formula SSSS) based on the fact that these species grow together in the Caucasian forests and can naturally hybridize with each other [3, 14, 18]. The same idea was supported by Rybin and Jukovsky, but later Georges Salesses questioned this hypothesis [18]. The origin of *P. domestica* remains somehow mysterious. There are three subspecies within *P. dome*stica: ssp. *insititia* (mirabelles and the so-called spilling), ssp. *oeconomica* (prunes), and ssp. *italica* (plums, reineclaudes, and all other kinds of plum fruits) [18].

For European plum, the plum fruits are round to oval, in different sizes and colors; the flesh is juicy, soft, and mostly clingstone; and the ripening time is earlier than those of prunes, but there are also some exceptions [16]. *Prunus domestica* is a very good source of genes for high sugar content, fruit flavor, late blooming, and high productivity but must be improved for frost and disease resistance (especially virus diseases) [14, 19].

Japanese plum has its origins in China (Yangtze River Basin), but for about 2000 years, it has been cultivated in Japan [1]. This species grew in warmer areas. It was imported to California in North America by Hough in 1870. The famous plum breeder Luther Burbank started his breeding in 1875 using all species available and produced thousands of seedlings and selected a lot of varieties [16]. The fruits are

Species	Common name	Origin	Chromosome number	Subspecies/ varieties
1	2	3	4	5
P. cerasifera Ehrh.	Cherry plum, Myrobalan	West Asia, Balkans (Serbia, Romania, Bulgaria, Greece), Caucasus	16 (24, 32, 48)	*atropurpurea*, *pissardi*, *pendula*, *elegans*, *divaricata*
P. cocomilia Ten.	Italian plum	Italy, Serbia	16	–
P. domestica L.	Garden plum, European plum	Europe, West Asia	48	–
P. insititia L.	Bullace, damson, mirabelle, reineclaude (gage plum)	Europa, West Asia	48	*subsylvestris*, *italica*, *syriaca*
P. monticola Koch	Taurus plum	Asia	16	–
P. salicina Lindl.	Japanese (Chinese) plum	China	16 (32)	–
P. simonii Carriere	Apricot plum, Simon plum	North China	16	*purpurea*
P. spinosa L.	Blackthorn, sloe	Europa, North Africa, West Asia	32 (16, 24, 48)	–
P. ussuriensis Kovalev and Kostina	Ussurian (Manchurian) plum	China	16	–
P. americana Marshall	Common wild plum	East USA, to the Rocky Mountains	16	*mollis*, *lanata*
P. angustifolia Marshall	Chickasaw plum	USA (New Jersey to the Florida); Illinois, Texas	16	*watsonii*, *varians*
P. hortulana L.H. Bailey	Hortulan plum	USA (Kentucky, Tennessee to the Iowa, Oklahoma, Texas, Louisiana); Alabama	16	*mineri*, *pubens*
P. maritima Marshall	Beach plum	USA (Virginia)	16	*flava* (with yellow fruit)
P. mexicana S. Wats.	Big-tree plum	USA, Mexico	16	*polyandra*, *fultonensis*
P. munsoniana Wight and Hedrick	Wild Goose plum	USA (Texas, Ohio, Kentucky)	16	–
P. nigra Aiton	Canadian plum	Canada, USA	16	–

Table 3.
Plum species with their common name, origin, and chromosome number. Sources: [5, 6, 19–21].

various, mostly large and firm with yellow base color overlaid by red and purple, very attractive, designated for the fresh market [1, 14].

4. Breeding objectives

Breeding new plum cultivars needs to anticipate the future requirements from the growers, market, and consumers.

A large number of breeding programs are developed in many countries from Europe (**Table 4**). Some of plum breeding programs have been reduced or stopped in the countries where production has declined or funding is no longer available. At the same time, some breeding programs became more private with less public funding [22].

The breeding objectives are general and specific. The general objectives include the following:

- Productivity
- Fruit quality
- Disease resistance, especially to plum pox virus.

The special objectives concern the following (**Table 3**):

- Late blooming and frost resistance in United Kingdom, Bulgaria, Moldova, and Belarus
- Winter hardiness in Latvia, Belarus, Moldova, Russia, Sweden, and Norway
- Short period of vegetation in Sweden
- Good storage in Norway
- Extended ripening period in Bulgaria and Romania
- Self-fertility in Latvia and Romania

Eighty percent of all breeding activities are carried on by *Prunus domestica* and only twenty percent by *Prunus salicina* [2, 13, 23–25].

5. Breeding methods

There are several breeding methods:

A. Conventional methods:

- Selection
- Hybridization
- Mutagenesis

B. Biotechnological methods:

- In vitro cells and tissue culture
- Induction of somaclonal variations
- Somatic hybridization
- Genetic engineering

DOI: http://dx.doi.org/10.5772/intechopen.92432

Country	Breeding centers	Breeder	Objectives
1	2	3	4
Romania	Research Institute for Fruit Growing, Pitesti Fruit Growing Research Station, Valcea Research Station for Fruit Growing, Bistrita	M. Butac M. Botu I. Zagrai	- Improvement of old cvs. Tuleu gras, Grase romanesti, Vinete romanesti - Fruit quality - PPV resistance - Ripening season extension - Self-fertility
Serbia	Fruit Research Institute, Cacak Faculty of Agronomy, Cacak	N. Milosevič I. Glisič T. Milosevič	- Improvement of old cv. Pozegača - Fruit size and quality - Ripening time - Resistance to PPV - Productivity
Germany	University of Hohenheim, Stuttgart Hochschule Geisenheim University	W. Hartmann M. Neumuller H. Jacob	- Fruit quality - PPV resistance by hypersensitivity
France	INRA, Bordeaux	R. Bernhard R. Renaud	- Fruit quality - PPV resistance by transgenic plants
Italy	University of Bologna University of Firenze University of Forli Private program	S. Sansavini E. Bellini V. Nancetti A. Liverani	Japanese and European plum breeding program: - Fruit quality - Fresh consumption - Resistance to PPV
United Kingdom	East Malling Research Station, Kent	K. Tobutt R. Jones T. Laxton	- Late blooming - Frost resistance - PPV resistance
Bulgaria	Fruit Growing Institute, Dryanovo Fruit Growing Institute, Troyan Fruit Growing Institute, Kyustendil Fruit Growing Institute, Plovdiv	V. Bozhkova A. Zhivondov	- Late blooming - Extended ripening period - Fruit quality - Resistance to PPV
Latvia	Latvia State Institute of Fruit Growing, Dobele	E. Kaufmane I. Gravite L Ikase	- Winter hardiness - Low vigor - Precocity - Fruit quality - Self-fertility - Early ripening
Belarus	Institute for Fruit Growing, Minsk	Z. Kazlouskaya V.A. Matveev M. Vasiljeva	- Fruit quality - Frost resistance
Republic of Moldova	Research Institute for Horticulture and Alimentary Technologies, Chisinau	M. Pintea, A. Juraveli	- Frost resistance - Fruit quality
Czech Republic	Research and Breeding Institute of Pomology, Holovousy	J. Blazek	- Fruit quality - Sharka tolerance
Poland	Research Institute of Horticulture, Skierniewice	T. Jakubowski E. Zurawicz J. Dominikowski E. Rozpara Z. Grzyb J. Szymanski	- Fruit quality - Productivity - PPV resistance
Russia	North-Caucasus Zonal Horticulture and Viticulture Research Institute, Krasnodar Krymsk Experiment Breeding Station,	R. S. Zaremuk E. M. Alekhina G. Eremin	- Winter hardiness - Fruit quality - Productivity

Country	Breeding centers	Breeder	Objectives
1	2	3	4
	Krymsk Far East Experiment Station, Pavlovsk, St. Petersburg Maicop Experiment Station Krasnodar Region Volgograd Experiment Station, Krasnoslobodsk		
Sweden	University of Agricultural Sciences, Balsgard	I. Hjalmarsson, V. Trajkovski	- Short period of vegetation - Fruit quality
Norway	Ullensvang Research Centre Division Njos, Lofthus	S.H. Hjeltnes	- Fruit quality - Good storage
Hungary	Research and Extension Centre for Fruit Growing, Ujfeherto	T. Lakatos	- Fruit quality - Processing

Sources: [3, 10, 26, 27].

Table 4.
Plum breeding centers and objectives in Europe.

Conventional breeding methods are still largely used in the majority plum breeding programs, and the most important of these are intra and interspecific hybridization and open pollination. These methods contributed in a substantially way to modify the genetic structure of quantitative traits of new plum cultivars and rootstocks [3, 10, 20, 28].

Intraspecific hybridization steps are parents' choice according to the breeding objectives, isolation and emasculation of flowers from the mother parent, collecting pollen from father parent, artificial pollination, control of fruits set, pick up of hybrids fruits, extracting the hybrids stones, and cultivation of seedlings.

Out of artificial sexual hybridization, several variants have been utilized: simple crossing, double crossing, pyramid type crossing, recessive crossing (back cross), and diallel crossing [29].

Between 1950 and 2000, most plum varieties have been created by intraspecific simple crossing (A × B): 'Stanley' ('d'Agen' × 'Grand Duke') in the USA; 'Valor' ('Imperial Epineuse' × 'Grand Duke') in Canada; 'Čačanska Lepotica' ('Wagenheim' × 'Pozegača'), 'Krina' ('Wagenheim' × 'Italian Plum') in Serbia; 'Jojo' ('Ortenauer' × 'Stanley') in Germany; 'Centenar' ('Tuleu gras' × 'Early Rivers') and 'Roman' ('Tuleu gras' × 'Early Rivers') in Romania, etc. [2, 29].

Using double crossing, in Romania, the 'Valcean' plum cultivar was obtained ['H8-12' ('Renclod Althan' × 'Wilhelmina Spath') × 'H 5-23' ('Renclod Althan' × 'Early Rivers')] [2, 29].

In the last 25 years, the pyramidal type cross [(A × B) × C] was more utilized.

Thus, in Romania, using this type of crossing, several varieties were registered: 'Minerva,' 'Sarmatic,' 'Pitestean' ['Tuleu timpuriu' ('Tuleu gras' × 'Early Rivers') × 'Early Rivers'] [2, 29].

In the last years, the interest for obtaining complex genotypes has increased, and for this reason, the interspecific hybridizations have been used.

Initially, interspecific hybridization was used to improve plum rootstocks. Thus, the plum rootstock 'Ishtara' is a complex interspecific hybrid between *P. domestica*, *P. cerasifera*, and *P. armeniaca* [30]. 'Jaspi' plum rootstock was obtained from the crossing between *P. salicina* Methley and *P. spinosa*. 'Marianna' rootstock was obtained from the combination between *P. cerasifera* and *P. munsoniana* [3].

Within the European FP 7 project, a breeding program for the creation of rootstocks with resistance to plum pox virus was carried out. Thus, at the Technical

University of Munich, the 'Dospina 235' (*P. domestica* × *P. spinosa*) and 'Docera 6' (*P. domestica* × *P. cerasifera*) rootstocks were obtained [31].

In Romania, there were crossed varieties belonging to the *P. domestica* and *P. insititia* species with the same number of chromosomes, and several cultivars were named: 'Silvia,' 'Ialomita,' 'Diana' ('Renclod Althan' × 'Early Rivers'), 'Renclod de Caransebes' ('Renclod Althan' × 'Wilhelmina Spath'), 'Doina,' 'Zamfira' ('Anna Spath' × 'Renclod Althan'), 'Romaner,' and 'Iulia' ('Tuleu gras' × 'Renclod Althan') [2, 6].

Hybridization between diploid species (*P. cerasifera*, *P. salicina*, *P. simonii*, *P. besseyi*, *P. americana*, *P. nigra*, *P. munsoniana*, *P. angustifolia*, and *P. hortulana*) can be very easy. For example, 'Santa Rosa' variety with American origin is a mixture between *P. salicina*, *P. simonii*, and *P. americana* [32].

In recent years, fruits of plum hybrids obtained from interspecific crosses have appeared on the world market:

- Interspecific hybrids between *P. domestica* and *P. armeniaca* called 'Plumcot®' (e.g., 'Red Velvet,' 'Royal Velvet,' 'Flavor Supreme,' 'Flavor Queen,' 'Rutland,' 'Plum Parfait,' 'Spring Satin,' and 'Yiksa'). In Bulgaria, Argir Zhivondov made crosses between *P. domestica* ('Stanley' cv). and *P. armeniaca* ('Modesto' cv.) and obtained the cultivar named 'Standesto' [21].

- Interspecific hybrids between (*P. domestica* × *P. armeniaca*) × *P. domestica* and (*P. salicina* × *P. armeniaca*) × *P. salicina*. The name of these hybrids is 'Pluot®'.

- Interspecific hybrids between (*P. domestica* × *P. armeniaca*) and *P. armeniaca* called 'Aprium®' (e.g., 'Flavor Delight,' 'Flavor Candy,' and 'Honey Rich Aprium').

Regarding clonal selections, these have been done into the old plum orchard from Romania with 'Vinete romanesti' cv., and the following clones were obtained: 'Vinete românești 300,' 'Vinete românești 303,' and 'Vinete românești 4.' Clonal selections produced also several rootstocks: 'Oteșani 8,' 'Oteșani 11,' and 'Voinești B' [7, 29].

Mutagenesis was applied also in Romania on buds and seeds, using gamma radiations (Co^{-60}) and X-rays (Roentgen). For examples, the Romanian plum cultivars 'Alina' and 'Tita' have been obtained through irradiation of 'Tuleu gras' seeds with X-rays [19, 29].

In recent years, genetic engineering and biotechnologies have an important role in plum breeding.

Thus, in vitro culture techniques are used to obtain rootstocks virus free. Also, the protocol for pollination and fertilization in vitro of some plum varieties was elaborated during the many years of investigations. For pollination in vitro, it is important to take the non-pollinated flowers at the stage of ovule receptivity for pollen tube. The ovules of 'Sweet Common Prune' were pollinated with pollen of 'Stanley.' Excised ovules were placed on white medium with 15% of sucrose and pollinated with pollen extracted from anthers. The fertilized ovules should be cultivated in the test tubes, in complete darkness, at a temperature of 25°C, for 60 days. After 7 days, percentage of growing embryos was determined [33–35].

Embryo culture is used to create varieties with very early ripening. Results in this direction had Gercheva and Zhivondov [36] taking immature embryos from the 'Burmosa' (*P. salicina*) and 'Ruth Gerstetter' (*P. domestica*) varieties and cultivating them on a culture medium Murashige and Skoog. Burmosa's embryos had a very good germination. In the last time, due to a very low germination percentage of some plum hybrid seeds in the breeding process, a new research using embryo culture method was started in Romania [36].

Molecular markers have a wide range of possible applications in plum breeding using markers such as RAPD and AFLPs [5]. In plum, gene transfer was applied especially for resistance to plum pox virus in order to create plum cultivars resistant to PPV. The PPV-CP (coat protein) gene was isolated, sequenced, cloned, and used for *Agrobacterium tumefaciens*—mediated transformation of plum [37]. Transgenic European plums were obtained that were grafted on the rootstocks of *P. domestica*, and their behavior at PPV was studied in the greenhouse. After 2 years of testing in greenhouse, a transgenic line C5 was resistant to PPV. This clone was registered as 'Honeysweet' cv. Later, this clone was also tested in the field, in Poland, Romania, and Spain under high-pressure infections and had a high level of PPV resistance [38–42].

6. Genetic sources of breeding

Germplasm collections are a major source of plant genetic diversity, which have an effect to improve crop. Collection, conservation, and evaluation of plant genetic resources are the most important conditions for breeding program. Plant breeders use these genetic resources in hybrid combinations because they are looking for new traits to be included into new varieties [1, 2, 43–47].

The existing ex situ collections can be a source of genes potentially useful as material in breeding work or sources of cultivars for a sustainable production.

In Europe there are a lot of plum genetic resources preserved in ex situ collections from about 30 countries.

A large number of plum accessions (estimated to be 4500) are kept in Russia and adjacent states, in four experimental stations located in different climatic conditions coordinated by Vavilov Institute of Plant Industry, Saint Petersburg. From this total, 2325 accessions (about 1600 original Russian cultivars) belong to *P. domestica* (2n = 48), and 2175 accessions are diploid plums (2n = 16), from which 500 genotypes belong to *P. cerasifera* [1].

Other European countries that have a large number of plum accessions are Belgium (616), Hungary (579), France (555), Italy (506), Bulgaria (377), the United Kingdom (380), Switzerland (326), Nordic countries (324), the Czech Republic (276), Portugal (263), Serbia (249), Turkey (232), Latvia (223), Germany (165), etc. [48, 49].

In Romania there are plum collections in two centers: RIFG Pitesti and UCv-SCDP Valcea. Ex situ conservation of accessions is done by different methods: conservation in the field collections (at RIFG Pitesti and UCv-SCDP Vâlcea), conservation in plastic containers, and cryoconservation, at −196°C, in liquid nitrogen (at UCv-SCDP Valcea) [50–53]. The plum cultivar collection at RIFG Pitesti, established in 1997, includes 550 accessions, and the plum rootstock collection, established in 2009, includes 92 accessions. The plum collections at SCDP Vâlcea, established in 1989, 1993, and 1996, include 361 accessions (species, cultivars, and rootstocks). So, from the total of 1003 accessions (cultivars and rootstocks), 34 are species and interspecific hybrids, 407 are local cultivars, 476 are foreign cultivars and rootstocks, and 86 are other genotypes (biotypes, hybrids, mutants, etc.). Most of the accessions belong to *Prunus domestica*, *Prunus insititia*, *Prunus cerasifera*, *Prunus spinosa*, and *Prunus salicina*.

The observations and measurements in ex situ collections were done according to the IBPGR *Prunus* descriptors updated by the ECP/GR *Prunus* Working Group members within the Genres CT95 No. 61 project titled "International network on *Prunus* genetic resources".

Most of these genetic resources have been evaluated and included in a European *Prunus* database (2254 accessions). In 1994, the European Cooperative Programme for Crop Genetic Resources Networks (ECP/GR) had the initiative to maintain this European *Prunus* database at the National Institute of Agronomic Research (INRA) Bordeaux, France, the manager being Emilie Balsamin [1].

From the ex situ collections, local varieties are very valuable, and many of them are characterized with good adaptability to environmental factors, high productivity, high resistance to pests and diseases, and fruits rich in nutrients necessary for human nutrition and different diets [54, 55].

The observations and determinations carried out over the years in the germplasm collections have revealed potential genitors with valuable traits for achieving the objectives from plum breeding programs.

Yield potential is a result of several factors, such as growth vigor, precocity, self-fertility, etc.

- *Growth vigor*. For, plum there are no sources of genes for low vigor or dwarf type. The following are sources of genes for low vigor: 'Stanley,' 'Marry Mather,' 'd'Ente 698,' 'Sugar,' 'Ealta Dorata di Coe,' 'Early Golden,' 'Belle de September,' 'Sticolase de Voinesti,' 'Prun negru,' 'Perje de toamna,' 'Galbene de Aninosani,' 'Grase de Becs,' 'Grase de Pesteana,' etc. [2, 6].

- *Precocity*. The growers require varieties with precocious. Good donors for precocity are 'Stanley,' 'Bluefree,' 'Čačanska Lepotica,' 'Čačanska Rodna,' 'Verity,' 'Valor,' 'Centenar,' 'Minerva,' etc. [6, 16].

- *Self-fertility*. Good donors for self-fertility are 'Stanley,' 'Anna Späth,' 'Standard,' 'Ontario,' 'Ialomiţa,' 'Diana,' 'Romanţa,' etc. [6, 16].

Frost resistance. The frost resistance of the plum varieties is in accordance with their origin and the place where they were created. Thus, the varieties originating in Western Europe have a low or medium resistance ('Reine Claude Verte' and 'd'Agen'), while those originating in North America and China are very resistant ('Opata,' 'Sapa,' and 'Waneta') [56]. The species *P. americana* and *P. nigra* transmit to the offspring the frost resistance [32]. *P. ussuriensis* species have a very high resistance to low temperatures, but its use in breeding is limited due to the fact that it blooms very early, a trait that is transmitted to the offspring; there is a risk of flower destruction if low temperatures occur during flowering [56]. The European varieties that have good resistances to frost are the following: 'Bonne de Bry,' 'Mount Royal,' 'Pozegača,' 'Stanley,' etc. Considering that the plum blooms early, immediately after the almond and apricot, the goal is to improve the resistance to the late frosts in the spring which, in recent years, have been more and more frequent. For example, in April 2009 and 2017 in Pitesti, Mărăcineni, Romania, temperatures of −1.7°C in the air and −3.4°C (2009) and −4.2°C (in 2017) in the soil were recorded. At these temperatures in 2009, the flowers of the varieties 'd'Agen' and 'Anna Späth' were destroyed in a proportion of 70–80%. In 2017, the low temperatures destroyed the young fruits. In this regard, the resistance of flowers to the late frosts can be achieved by creating late-flowering varieties. The late-flowering varieties that transmit this trait to the descendants can be mentioned: 'De Bistriţa,' 'Vinete româneşti 300,' 'Bistriţene de Haţeg,' 'Prune roşii,' 'Busuioace de Geoagiu,' 'Perje de toamnă,' 'Tuleu timpuriu,' 'Pescăruş,' 'Tămâioasă de Bistriţa,' 'Albe de Bilceşti,' 'Superb,' 'Prune roşii,' 'Prun de stepă,' 'Tuleu gras,' 'Pozegača,' 'd'Agen,' 'Late d'Agen,' 'Drjanovska Sliva,' 'Korai Besztercei,' 'Troianskaia Sliva,' 'Hamanova Svetska,' 'R.C. Violet,' 'OK,' 'Belle de Liege,' 'Belle de Louvain,' 'Mohawk,' 'R.C. d'Oullins,' etc. [6, 53, 57].

Ripening time. In the plum breeding program, it is desirable to extend the ripening season, creating very early varieties, but also very late. Donors for earliness are recommended: 'Ersinger,' 'Petrovača' [3], 'Early Rivers,' 'Ruth Gerstetter,' 'Čačanska Lepotiča,' 'Čačanska Rana,' 'Ialomiţa,' 'Diana,' 'Scolduş de vară,' 'Boboloase,' etc. [2, 6]. Potential genitors for lateness are: 'Grand Duke' [58], 'Anna Späth,' 'President,' 'Record,' 'Vinete româneşti,' 'Grase româneşti,' etc. [2, 6, 57].

Fruit characters. Fruit quality represents an important breeding objective, as the consumer requirements are constantly increasing [59–62]. The new variety will be successful only if certain characteristics are met, such as fruit size, skin color, commercial aspect, sweetness, acidity, firmness, flavor, and juiciness [26, 63–65].

- *Fruit size.* Generally, regarding this trait, the most influence on the progenies is exercised by the mother parent and the cumulative effect of both parents [6]. Therefore, in order to obtain hybrids with large fruits, it is desirable for parents to have large fruits. At the European plum, the fruit size is absolutely necessary, especially since most varieties with special tasting qualities have small fruits (e.g., 'Tuleu gras'). They are of interest as genitors of fruit size for the following varieties: 'Jubileum' from Sweden; 'Tophit'; 'Haganta' from Germany [16]; 'Valor' and 'Vision' from Canada; 'Čacčanska Najbolja' from Serbia; and 'Record,' 'Vâlcean,' 'Tita,' 'Carpatin,' and 'Romanţa' from Romania [6, 57].

- *Fruit shape*. At *Prunus domestica* species, the ellipsoidal fruit shape is dominant over the spherical one, whereas at *Prunus salicina* there is no dominance for the spherical or elongated shape [32]. On the fruit market in Middle Europe elongated shape fruits are preferred. The following varieties can be used as parents for the elongated shape: 'Stanley,' 'Pozegača,' 'Vinete româneşti,' 'Tuleu gras,' 'Centenar,' etc. [6]. The following varieties can be used as parents for the spherical shape: 'Anna Späth,' 'Kirke,' 'California Blue,' Gras ameliorat,' 'Grase de Becs,' etc. [6, 57].

- *Fruit color*. For fresh market, the fruit color is an important trait. The fruit color of plums ranges from dark blue to blue, purple, red and yellow. At *P. domestica* and *P. salicina* species, the skin color is determined by an allelic series, which the allele for the yellow color is recessive to the allele for the blue, red, or purple colors [32]. In the European fruit market, consumers prefer two skin colors: green with different shades (France and parts of Germany) and blue in most countries. The following can be used as parents for the blue color of the skin: 'Stanley,' 'Standard,' 'Oneida,' 'Valor,' 'Vision,' 'Kirke,' 'Hackmann,' 'Bluebell,' 'Bluefree,' 'Negre de Seini,' 'Negre de Bilceşti,' 'Vinete româneşti,' 'Tămâioasă de Bistriţa,' 'Piteştean,' 'Pescăruş,' 'Centenar,' 'Dâmboviţa,' etc. [6, 53, 57].

- *Fruit taste*. The taste is the most important aspect of fruit quality. The fruit taste, expressed as a ratio between the main components (sugar, acidity, tannins, vitamins, and aroma) has a great variability, according to the consumers' requirements but also on the origin of the varieties. For example, people from Southern Europe and those from Asia prefer sweet fruits. In other countries, people prefer fruit with a balance between sugar and acidity [16]. Also, a balanced taste has varieties originating from *P. domestica* and *P. insititia* [6, 57]. Thus, the following varieties are recommended as genitors for good and very good taste: 'Tuleu gras,' 'Centenar,' 'Gras ameliorat,' 'Grase de Becs,' 'Uriașe de Sibiu,' 'Agent,' and 'Andreea' in Romania [6]; 'Bijelica sitna' and 'Prskulja' in Bosnia and Herzegovina; 'Moravka,' 'Metlaš,' 'Obični piskavac,' and 'Čačanska Najbolja' in Serbia; 'Auerbacher,' 'Ortenauer,' and 'Wangenheims' in Germany; 'Italian prune,' 'President' in the USA, and 'd'Agen' in California and France [3].

Sharka (PPV) resistance. One of the major objectives in plum breeding, both in our country and worldwide, is the resistance to viral diseases, especially to plum pox virus, one of the most damaging pathogens causing yield losses over 70%, especially to susceptible cultivars [66, 67]. The yield of sensitive varieties, which externalize disease's symptoms on fruit, loses commercial value for fresh market. The fruits affected are blemished, misshapen, and distorted with sunken lesions in flesh. If don't drop prematurely and rich harvest maturity, the fruits have poor flavor, small size, low sugar content and anthocyanin. These fruits can be sold at low price, only to distilleries for brandy processing [67].

The only efficient way to fight against this virus is to develop resistant or tolerant cultivars. The following can be used as a gene source for the resistance to plum pox virus (Sharka): 'Popesti,' 'Vieneze,' 'Roman,' 'Cisnadie,' 'Lachi,' 'Alina,' 'Ungurești,' 'Ciorași de Prahova,' 'Flora,' 'Negre de Saru,' 'Negre de Bilcești,' 'Krimska,' 'Peche,' 'Chabat,' 'Hüttner,' 'Belle de Liege,' 'Kirke,' 'Marry Mather,' 'Giant plum,' and 'Condata' [53].

In Germany, at Hohenheim University, Stuttgart, Professor Hartmann achieved very good results on the PPV resistance line, surprisingly, by using the 'Stanley' variety as a parent. An "absolute resistance" through hypersensitivity (HR—hypersensitive reaction) was identified in the 'Jojo' variety obtained from the cross between the 'Ortenauer' and 'Stanley' varieties [16, 68].

Zagrai et al. [42] showed that transgenic plants, with incorporated coatprotein (CP) genes, through mediation with the bacterium Agrobacterium tumefaciens, confer protection against Sharka [42]. A transgenic clone, C5 (named 'Honeysweet'), was also studied under the conditions of different countries from Europe, remaining free of PPV after years of testing [42].

7. Achievements and prospects

A large number of breeding programs are developed in different countries from Europe. Some of the plum breeding programs have been reduced or stopped in the countries where production has declined or funding is no longer available. At the same time, some breeding programs became more private with less public funding [10, 22].

In *Romania*, during 60 years of breeding work, over 2,000,000 plum flowers were pollinated at different centers: Pitești, Vâlcea, Bistrița, and Strejești. From the hybrid seeds, approximately a thousand hybrids were obtained which, thanks to their variability, enabled the selection of many new valuable hybrids. As a result 40 new plum cultivars were registered. In the first two breeding stages (1955–1970), the cultivars 'Tuleu timpuriu' (1967, 'Tuleu gras' × 'Peche'), 'Superb' (1968, 'Tuleu gras' × 'Abbaye d'Arton'), 'Gras ameliorat' ('Grase românești'—self-pollination), 'Vinete românești 300' (1970, 'Vinete românești'—selection), and 'Tuleu dulce' (1968, 'Tuleu gras' × 'd'Agen') were selected. Among the basic genitors, 'Tuleu gras' cv. was by far the best, giving the greatest number of promising selections. During this period, the cultivars 'Renclod Althan,' 'd'Agen,' 'Early Rivers,' and 'Wilhelmina Späth,' used as character genitors, proved of particular value. In the third breeding stage (1970–1980), the cultivars 'Centenar' (1978, 'Tuleu gras' × 'Early Rivers'), 'Silvia' (1978, 'Renclod Althan' × 'Early Rivers'), 'Albatros' (1979, 'Tuleu gras'—open pollination), and 'Pescarus' (1979, 'Renclod Althan' × 'Wilhelmina Späth') were registered. This was a very fruitful stage, because the experimental plots and field trials within the national research network gave the possibility to both select the new autochthonous cultivars and also to choose the adequate area for testing each of them. In the fourth stage (1980–1990), the following varieties were registered: 'Ialomița' (1981, 'Renclod Althan' × 'Early Rivers'), 'Diana' (1981, 'Renclod Althan' × 'Early Rivers'), 'Piteștean' (1981, 'Tuleu timpuriu' × 'Early Rivers'), 'Carpatin' (1981, 'Tuleu gras' × 'Early Rivers'), 'Dambovița' (1981, 'Tuleu gras' × 'Anna Späth'), 'Record' (1982, 'Renclod Violet'—open pollination), 'Minerva' (1984, 'Tuleu timpuriu' × 'Early Rivers'), 'Flora' (1989, 'Tuleu gras' × 'Renclod Violet'), and 'Sarmatic' (1989, 'Tuleu timpuriu' × 'Early Rivers'), with large fruit and high yields. In the last stage (after 1990), which is still going on, the greatest number of cultivars was recorded: 'Renclod de Caransebeș' (1990, 'Renclod Althan' × 'Wilhelmina Späth'), 'Vâlcean' (1990, 'H 8/12' × 'H 5/23'), 'Bărăgan 17' (1990, 'Tuleu gras' × 'Early Rivers'), 'Tita' (1991, 'Tuleu gras'—irradiated stones), 'Alina' (1991, 'Tuleu gras'—irradiated stones), 'Andreea' (2000, 'H 27/87' open pollination), 'Delia' (2002, 'Vanat Italy' × 'Anna Späth'), 'Iulia' (2002, 'Tuleu gras' × 'Renclod Althan'), 'Ivan' (2003, 'Tuleu gras' × 'Vanat Italia'), 'Jubileu 50' (2003, 'Tuleu gras' × 'De Bistrița'), 'Dani' (2004, 'Tuleu gras' × 'Grase românești'), 'Doina' (2004, 'Anna Späth' × 'Renclod Althan'), 'Geta' (2004, 'Centenar' × 'Ialomița'), 'Matilda' (2004, 'Anna Späth' × 'd'Agen'—irradiated with Co^{60}), 'Roman' (2004, 'Tuleu gras' × 'Early Rivers'), 'Agent' (2004, selection within a population of seedlings resulted from open pollination), 'Romaner' (2005, 'Tuleu gras' × 'Renclod Althan'), 'Zamfira' (2005, 'Anna Späth' × 'Renclod Althan'), 'Elena' (2005, 'Tuleu gras' × 'Stanley'), 'Alutus' [2010 ('R.C. Althan' × 'Early Rivers') × ('R.C. Althan' × 'Wilhelmina Späth') × mixed pollen], 'Topval' (2010, 'Tuleu gras' × 'Stanley'), and 'Romanța' (2012, 'Stanley' × 'Vâlcean'). Some of these cultivars are proven to be tolerant to PPV besides their high-quality fruit and yields [19].

In *Serbia*, about 530 hybrid combinations were made in the European plum breeding program starting in 1949 at the Fruit Research Institute, Čačak, of which around 30,000 hybrids have been produced and 16 varieties have been released. The oldest varieties obtained are 'Čačanska Rodna' (1975, 'Stanley' × 'Požegača'), 'Čačanska Rana' (1975, 'Wangeiheim' × 'Požegača'), 'Čačanska Secer' (1975, 'd'Agen' × 'Pacific'), 'Čačanska Lepotica' (1975, 'Wangeiheim' × 'Požegača'), 'Čačanska Najbolja' (1975, 'Wangeiheim' × 'Požegača'), 'Jelica' (1986, 'Požegača' × 'California Blue'), 'Valerija' (1986, 'Hall' × 'Ruth Gerstetter'), and 'Valjevka' (1986, 'd'Agen 707' × 'Stanley'). After 2000 several promising new cultivars with high commercial potential were named: 'Mildora' (2004, 'Large Sugar' × 'Čačanska Lepotica'), 'Boranka' (2004, 'California Blue' × 'Ruth Gerstetter'), 'Timočanka' (2004, 'Stanley' × 'California Blue'), 'Krina' (2005, 'Wangenheim' × 'Italian Plum'), 'Pozna Plava' (2008, 'Čačanska Najbolja'—self-pollination), 'Zlatka' (2008, 'Large Sugar' × 'Zuta Boutilcovidna'), 'Nada' (2012, 'Stanley' × 'Scolduș'), 'Petra' (2018, 'Stanley' × 'Opal'), and 'Divna' (2018, 'Stanley' × 'Čačanska Rana'). Cultivars such as 'Čačanska Lepotica,' 'Čačanska Najbolja,' and 'Čačanska Rodna' were used as parents in different breeding programs in some European countries, especially Germany, Bulgaria, the Czech Republic, and Romania [3, 69–71].

In *Germany*, plum breeding is carried out in two centers: the University of Hohenheim, Stuttgart, and Geisenheim Research Station. In the University of Hohenheim in Stuttgart, new varieties were created by Professor Walter Hartmann, valuable through resistance to plum pox virus, fruit taste, attractive appearance, and multiple uses of the fruit. These varieties are 'Hanita' (1991, 'President' × 'Auerbacher'), 'Katinka' (1992, 'Ortenauer' × 'Ruth Gerstetter'), 'Elena' (1993, 'Italian Prune' × 'Stanley'), 'Jojo' (1993, 'Ortenauer' × 'Stanley')—the first plum cultivar immune to the plum pox virus—'Felsina' (1994, 'Italian Prune' × 'Ersinger'), 'Tipala' (1995, 'Tiroler Zuckerzwetsche' × 'Opal'), 'Tegera' (1995, 'Ortenauer' × 'Ruth Gerstetter'), 'Presenta' (1996, 'Ortenauer' × 'President'), 'Colora' (2003, 'Ortenauer' × 'Ruth Gerstetter'), 'Azura' (2003, 'Hanita' × 'Čačanska Lepotiča'), 'Haganta' (2003, 'Čačanska Najbolja' × 'Valor'), 'Haroma' [2003, ('Ortenauer' × 'Stanley 34') × 'Hanita'], 'Habella' (2003, 'Ortenauer' × 'Stanley 34'), Hanka ('Hanita' × 'Katinka'), 'Juna' ('Katinka' × 'Zwintschers'), 'Jofela' (2013, 'Jojo' 'Felsina'), 'Jolina' (2013, 'Jojo' × 'Haganta'), and 'Joganta' (2014, 'Jojo' × 'Haganta') [3, 6, 72, 73]. At the Research Station of Fruit Growing, Geisenheim, Professor Helmut Jacob carried out an extensive breeding program for 25 years, which resulted in 12 plum varieties, of which 10 for fresh consumption and 2 for distillation [74]. The plums designated for fresh consumption are: 'Topfive' (1987, 'Čačanska Najbolja' × 'Auerbacher'), 'Topper' (1988, 'Čačanska Najbolja' × 'Auerbacher'), 'Topking' (1988, 'Čačanska Najbolja' × 'Italian Prune'), 'Tophit Plus' (1988, 'Čačanska Najbolja' × 'President'), 'Top 2000' (1991, 'Stanley' × 'NN'), 'Topfirst' (1993, 'Čačanska Najbolja' × 'Ruth Gerstetter'), 'Topstar Plus' (1993, 'Ersinger' × 'Čačanska Najbolja'), 'Toptaste' (1994, 'Valor' × 'German Prune'), 'Topgigant Plus' (1994, 'Čačanska Najbolja' × 'President'), and 'Topend Plus' (1994, 'Čačanska Najbolja' × 'Valor'). The two varieties destined for distillation are 'Bellamira' (1994, 'Čačanska Najbolja' × 'Mirabelle from Nancy') and 'Miragrande' (1995, 'Mirabelle Herrenhausen' × 'Yellow Plum') [75].

In *Bulgaria*, during the 60-year period of the breeding program, 29 plum cultivars were created in centers such as Dryanovo, Troyan, Kyustendil, and Plovdiv. By improving the old Kjustendilska cultivar, breeders like Marinov, Vitanov, Balev, Minev, Enev, Videnov, Georgiev, and Velkov obtained the following varieties: 'Dryanovska,' 'Sinya Yubileyna,' 'Gulyaeva,' 'Gabrovska,' 'Pop Hariton,' 'Burya,' 'Strinava,' 'Nevena,' 'Vitanova,' 'Edra trankosliva,' 'Baleva sliva,' 'Strumska sinya,' 'Kyustendilska ranna,' 'Izobilie,' 'Kyustendilska krasavitsa,' and 'Plovdivska.' After 2000, breeders Zhivondov and Bozhkova created our plum varieties with tolerance to plum pox virus and resistance to late frost and drought: 'Plovdivska renkloda' and 'Sineva' ('Stanley'—open pollination), 'Ulpia' ('President'—open pollination), and 'Ostromila' ('Pacific' × 'Serdika 2') [76].

In *France*, the plum breeding program started in 1947 at INRA Bordeaux, and eight plum varieties (6 varieties for drying (prunes) and 2 varieties for fresh consumption (plums)) were registered: 'Primacotes®Coten' (1982, 'd'Ente' × 'd'Ente 629'), 'Tardicotes®Enduke' (1982, 'd'Ente' × 'Grand Duke'), 'Lorida®Enspa' (1985, 'd'Ente' × 'Anna Späth'), 'Lemburn' and 'Double Robe' (1985, 'd'Ente' mutagenesis), 'Spurdente®Ferco' (1986, 'd'Ente 707' mutagenesis), 'Fermareine®Bellina' (1993, 'Reine Claude Verte' × 'Reine Claude de Bavay'), and 'Ferbleue®' (1985, 'Reine Claude Verte' × 'California Blue') [3, 6, 77, 78]. French plum breeding program has stopped.

In *England*, many of the varieties created by Thomas Rivers are widespread today in culture: 'Czar,' 'Victoria,' 'Early Prolific,' 'Early Transparent,' 'Golden Transparent,' 'Heron,' 'Monarh,' 'President,' 'Early Rivers,' and 'Blue Tit.' 'President' and 'Early Rivers' cultivars are used as parents in breeding work from some

European countries. Thomas Laxton has created varieties such as 'Laxton Beautiful' and 'Laxton Cropper.' At the East Malling Research Station, 'Cox's Emperor' and 'Yellow Egg' varieties were obtained but also other more recent varieties: 'Avalon' (1990, 'Reeves'—open pollination), 'Excalibur' (1990, 'Cox's Emperor'—open pollination), and 'Guinevere' (obtained in 2000)—these three varieties being created to replace the old Victoria' cultivar. England plum breeding program has stopped [3, 6].

In *Italy*, there are plum breeding programs both in public (at Universities Bologna, Firenze and Forli) and in private, and the following varieties have been obtained: 'Parlantina' (1987), 'Grossa di Felisio' (1988), 'Big Egg' (1989), 'Firenze 90' (1990), 'Grossa di Solarolo,' 'Empress,' 'Sugar Top' (2000, 'Susino II' × 'Stanley'), 'Prugna 29' (2000, 'French improved' × 'Stanley'), 'Liablu' (2000, 'Empress' × 'Ruth Gerstetter'), and 'Maria Novella' (2000) [6, 10, 79].

In the present, in *Russia*, there are 12 plum breeding programs. After 2000, 44 new plum varieties were released: 'Aleksiy,' 'Bolhovcanka,' 'Kantemyrovka' (at the All Russian Research Institute of Fruit Crop Breeding, Orel), 'Ballada,' 'Osenniy Suvenir,' 'Prikubanskaya,' 'Debjut,' 'Golubaja Mechta,' 'Milena,' 'Podruga,' 'Gertsog,' 'Krasotka,' 'Charodeyka,' 'Osennyaya,' 'Krasnodarskaya,' 'Beglyanka' (at the North Caucasian Zonal Research Institute of Horticulture and Viticulture Krasnodar), Vengerka Korneevska, Nižegorodkaya, Renklod Scherbinskiy (at Nizhny-Volzhsky Research Institute of Agriculture, Moscow), Vecherniy Zvon, Viola, Galateya, Pamyat Finayeva (at the Samara Zonal Experimental Station of Horticulture, Samara), 'Volžanka,' 'Kazanskaya,' 'Pamyat Hasanova,' 'Rakitovaya,' 'Teknovskaya Golubka' (at Tatar Scientific Research Institute of Agriculture, Kazan), 'Zanyatnaya,' 'Syniy Dar,' 'Suhanovskaya,' 'Utro,' 'Yahontovaya,' 'Kantemirovka' (at the All Russian Horticultural Institute of Breeding, Agrotechnology and Nursery, Moscow), 'Indira' (at the Kuibyshev Experimental Station for Horticulture Samara), 'Predgornaya,' 'Sverhrannyaya' (at the Dagestan Breeding Experimental Station of fruit and Berry Crops, Buynaksk), 'Antonina' (at the Maritime Fruit and Berry Experimental Station of the Maritime Research Institute of Agriculture in Vladivostok), 'Startovaya,' 'Zarecnaya Rannaya,' 'Renklod Kursakova,' 'Nochka' (at the All Russian Research Institute of Genetics and Breeding of Fruit Plants in Michurinsk), 'Viktorina,' and 'Baykalskaya' (at the Nikitsky Botanical Garden—National Science Center Yalta and the Scientific Research Institute of Horticulture of Siberia, Barnaul) [3, 80–83].

In *Belarus*, at the Institute for Fruit Growing, Minsk, the following varieties were created: 'Dalikatnaya' (2005, 'Evrasia 21' × 'd'Agen'), 'Kroman' (2005, 'Perdigron' × 'd'Agen'), 'Charadejka' ('Doneckaya rannayaya' × 'Victor'), 'Narach' (2008, 'd'Agen' × 'Renclod Althan'), 'Venera' (2009, 'Narach' × 'Wangenheim Fruhzwetsche'), 'Ranaya Losickaya', 'Vengerka Belorusskaya' (2010, 'Stanley' × 'Dalikatnaya'), 'Volat', and 'Nagrada Nemanskaya' [10, 84, 85].

Important achievements in the plum breeding are also in the *Republic of Moldova*, the work being started by Hramov in 1946 and continued by Condratiev, Juraveli, and Levițcaia at the Research Institute for Horticulture and Alimentary Technologies, Chisinau. Of these we mention 'Chisinovskaia Raniaia,' 'Vengherka Krupnaia Sladkaia,' 'Vengherka Jubileinaia,' 'Sopernița,' 'Raniaia Hramova,' 'Udlinionaia,' 'Pamiati Kostinoi,' 'Pozdniaia Hramova,' 'Pamiati Vavilova' (2010, 'Reine Claude Verte' × 'Vanat Italy'), 'Crasa Oseni' (2010, 'Reine Claude Verte' × 'Ispolinscaia'), 'Superpresident,' 'Ajur 1,' 'Naiboleco,' and 'Cabardinca' [3, 6, 10].

Plum breeding in *Estonia*, started in 1945 by Julius Eslon at the Polli Horticultural Institute, has had four stages so far, the varieties registered being the following: 'Karski,' 'Polli Munaploom,' 'Suhkruploom,' and 'Polli Varane,' created in the first stage; 'Julius,' 'Esloni Varane,' and 'Polli Viljakas,' obtained in the second stage; 'Amitar,' 'Ave,' 'Kadri,' 'Liisu Norgen,' 'Sargen,' 'Vilkana,' 'Vilmitar,' and 'Vilnor,' created in the third stage, from 1964 to 1985); and 'Kaidi,' a variety created in the last stage starting in 1986 [86, 87].

In *Latvia*, the breeding program carried out at the Latvia State Institute of Fruit Growing, Dobele, based on a large germplasm fund (184 plum accessions, of which 25 are indigenous varieties), contributed to the completion of the plum assortment with 12 new varieties: 'Agra Dzeltena' (now is growing only as pollinator), 'Inese,' 'Minjona,' 'Zemgale,' 'Lase,' 'Ance' (2010, 'Jubileum' × 'Opal'), 'Adele' (sin. 'Adelyn'; 2010, 'Laxton's Early' × 'Ruth Gerstetter'), 'Lotte' (2010 ['Jubileum' × 'BPr 5613' ('R.C. Reforma' × 'Ruth Gerstetter')], 'Sonora' (2010, 'La Crescent' × 'Jefferson'), 'Zane,' 'Laine,' and 'Lienite' [10, 88, 89].

In *Poland*, plum breeding program started at the Research Institute of Pomology and Floriculture at Skierniewice in the 1960s, and seven new cultivars were registered: 'Kalipso,' 'Emper,' and 'Polinka' (obtained by control hybridization); 'Dabrowice Prune' (seedling of the common plum—'Wegierka Zwykla'—from open pollination); 'Nectavit,' 'Tolerant'; and 'Promis' (clones of the common plum—'Wegierka Zwykla') [90].

In the *Czech Republic*, at the Research Breeding Institute of Pomology, Holovousy, 13 new plum cultivars were released: 'Amátka' (2011, 'Čačanska Lepotica' × 'Gabrovska'), 'Dwarf' (2011, 'Asatan' × 'Čačanska Najbolja'), 'Staňa' (2011, 'Stanley' × 'Gabrovska'), 'Kamir,' 'Hololepa,' 'Simona,' 'Samera,' 'Lipniská,' 'Malenovická,' 'Paní Háje,' 'Těchobuzická,' 'Valentýnka,' and 'Vítek' [91].

In *Sweden* and *Norway*, countries with a climate less favorable to plum culture, the results of the breeding programs were 'Edda' (1980, 'Czar' × 'Peche') in Norway and 'Opal' (1925, 'Oullins Gage' × 'Early favorites'), 'Herman' (1983, 'Czar' × 'Ruth Gerstetter'), 'Ariel' (1983, 'd'Autumn Compot' × 'R.C. Althan'), 'Jubileum' (1985), 'Madeleine' (1987, 'Hackmann' × 'Victoria'), 'Emil' (1989), and 'Violetta' (1991, 'Grand Duke' × 'Herman') in Sweden [6].

In *Canada*, at present, the plum breeding program has been suspended. However, from the old breeding program, starting in 1913 at the Horticultural Research Institute of Ontario in Vineland, 10 European plum cultivars were named: 'Valor' and 'Verity' (1967, 'Imperial Epineuse' × 'Grand Duke'), 'Vision' (1967, 'Pacific' × 'Albion'), 'Veeblue' (1984, 'Imperial Epineuse' × 'President'), 'Voyageur' (1987, 'Ruth Gerstetter'—open pollination), 'Victory' (1992, 'Vision' × 'Valor'), Valerie ('Valor' × 'California Blue'), 'Vanette' ('Early Rivers' × 'Stanley'), 'Vibrant' ('Valor' × 'California Blue'), and 'Violette' ('Verity' × 'Bluebell') [92, 93].

In the *USA*, at the New York Agricultural Experiment Station, Geneva, several cultivars have been released: 'Hall' (1923), 'American Mirabelle' (1925), 'Stanley' (1926, 'd'Agen' × 'Grand Duke'), 'Albion' (1929), 'Iroquois' (1966, 'Prugna di Italy' × 'Hall'), 'Mohawk' (1966), 'Oneida' (1966), 'Seneca' (1972), 'Moyer Prune' (1984), 'Castleton' ('Valor' × 'Iroquois'), 'Longjohn' ('Iroquois' × 'CA4A33L'), 'Polly' ('Oneida'—self -pollination), 'Kenmore' ('Standard' × 'Stanley'), and 'Amers' ('Stanley' × 'Standard'); the last five cultivars are registered after 2000. Among them, 'Stanley' cv. is the most widely grown cultivar worldwide [6]. At the Missouri State Fruit Experiment Station, in 1947, three European plum cultivars were released: 'Radiance,' 'Bluebell,' and 'Bluefree' ('Stanley' × 'President'). At the University of California at Davis, cultivars for dessert use were released ('Emperor' cv.), but also cultivars suitable for drying ('Sutter', 'Tulare Giant' in 2000, and 'Muir Beauty' in 2005) [9, 94]. Other European plums released in the USA at Oregon Agricultural Experiment Station are 'Gardner' (1923) and 'Hildreth' (1982) [9].

Recently, at the USDA Kearneysville WV, the researchers are focused on creating plum varieties with resistance to PPV using gene manipulation methods, and the first transgenic cultivar—'Honey Sweet'—was registered, which included a gene from the virus coat protein [39].

Author details

Madalina Butac
Research Institute for Fruit Growing, Pitesti, Romania

*Address all correspondence to: madalinabutac@yahoo.com

References

[1] Blažec J. A survey of the genetic resources used in plum breeding. Acta Horticulturae. 2007;**734**:31-45

[2] Butac M, Botu M, Militaru M, Mazilu CR, Duțu I, Nicolae S. Plum germplasm and breeding in Romania. Proceedings of the Latvian Academy of Sciences: Section B. 2019;**73**(3):214-219. DOI: 10.2478/prolas-2019-0034. 720

[3] Milosevič T, Milosevič N. Plum (*Prunus* spp.) breeding. In: Advances in Plant Breeding Strategies: Fruits. Vol. 2018. Verlag, New York: Springer International Publishing AG, part of Springer Nature; 2018. pp. 162-215. Available from: https://www.springer.com/gp/book/9783319919430

[4] Pirkhezri M, Fatahi Mogadam MR, Ebadi E, Hassani D, Abdoosi V. Morphopomological study of some new Japanese plum (*Prunus salicina* Lindl) cultivars grown in Iran. International Journal of Biosciences. 2014;**5**(8):180

[5] Topp BL, Russel DM, Neumüller M, Dalbo MA, Liu W. Plum. Chapter 15. In: Badenes ML, Byrne DH, editors. Fruit Breeding. Springer; 2012. pp. 571-621

[6] Butac M. *Ameliorarea prunului* (Plum Breeding). Universitatii din Pitesti; 2010. 101 pp. (in Romanian)

[7] Botu I, Botu M. Limits and perspectives in plum cultivar breeding using conventional methods. Acta Horticulturae. 2007;**734**:321-325

[8] Okie WR, Weinberger JH. Plums. In: Janick J, Moor JN, editors. Fruit Breeding, vol. I, Tree and Tropical Fruits. New York: John Wiley and Sons, Inc; 1996. pp. 559-607

[9] Okie WR, Ramming DW. Plum breeding worldwide. HortTechnology. 1999;**9**(2):162-176

[10] Butac M, Bozhkova V, Zhivondov A, Milosevic N, Bellini E, Nencetti V, et al. Overview of plum breeding in Europe. Acta Horticulturae. 2013;**981**:91-98

[11] FAO—Food and Agriculture Organization of the United Nations. Data Base. 2019. Available from: http://www.fao.org/faostat/en/#data/QC

[12] Potter D, Eriksson T, Evans RC. Phylogeny and classification of Rosaceae. Plant Systematics and Evolution. 2007;**266**(1-2):5-46

[13] Butac M, Zagrai I, Botu M. Breeding of new plum cultivars in Romania. Acta Horticulturae. 2010;**874**:51-58

[14] Butac M, Militaru M, Chitu E, Plopa C, Sumedrea M, Sumedrea D. Differences and similarities between some European and Japanese plum varieties. Acta Horticulturae. 2019;**1260**: 129-136. DOI: 10.17660/ActaHortic.2019.1260.21

[15] Goulao L, Monte-Corvo L, Oliveira M. Phenotic characterization of plum cultivars by high multiplex ratio markers: amplified fragment length polymorphisms and inter-simple sequence repeats. Journal of the American Society for Horticultural Science. 2001;**126**(1):72-77

[16] Hartmann W, Neumüller M. Plum breeding. In: Breeding Plantation Tree Crops—Temperate Species. Verlag, New York: Springer on line; 2009. pp. 161-231. Available from: https://www.springer.com/gp/book/9780387712024#reviews

[17] Crane MB, Lawrence WJC. The Genetics of Garden Plants. London, UK: MacMillan and Co., Ltd.; 1934

[18] Neumüller M. Fundamental and applied aspects of plum (*Prunus*

domestica) breeding. In: Fruit, Vegetable and Cereal Science and Biotechnology. Global Science Books; 2011. pp. 139-156

[19] Cociu V, Botu I, Minoiu N, Pasc I, Modoran I. Prunul (Plum). Romania: Conphys; 1997. pp. 46-51. (in Romanian)

[20] Botu I, Botu M, Papachatzis A, Cosmulescu S, Preda S. Evolution of plum culture: Constrains and perspectives. Acta Horticulturae. 2012; **968**:19-24

[21] Zhivondov A. Plumcots—Remote Hybridization of Fruit Species in Bulgaria. Bulgaria: Plovdiv; 2012

[22] Hartmann W. Opening address ninth international symposium on plum and prune genetics, breeding and pomology, Palermo, Italy. Acta Horticulturae. 2010;**874**:19-24

[23] Bellini E, Nencetti V, Natarelli L. New selections of yellow Japanese plum obtained in Florence at the Dofi. Acta Horticulturae. 2010;**874**:321-326

[24] Blažek J, Pistekova I. The first results from evaluation of plum cultivars grown in a very dense planting. Acta Horticulturae. 2010;**968**:99-109

[25] Jacob H. Plum breeding worldwide. I. Symposium on plum of Serbia. In: Book of Abstracts. 2006. p. 15

[26] Blažek J, Paprstein F. Development of fruit quality within top apple cultivars based on the consumer preference testing in last 34 years. Horticultural Science (Prague). 2014; **41**(1):10-18

[27] Milosevič N, Glišić I, Lukić M, Popović B, ĐorĎević M. Plum breeding in the Fruit Research Institute of Čačak, Serbia—Results of the last 15 years. Acta Horticulturae. 2019;**1260**:29-34

[28] Carrasco B, Meisel l, Gebauer M. Breeding in peach, cherry and plum: From a tissue culture, genetic, transcriptomic and genomic perspective. Biological Research. 2013; **46**(3):219-230

[29] Botu M, Achim G, Botu I, Preda S, Vicol A. Genetic gain achieved in plum breeding programs in Romania. Acta Horticulturae. 2012;**968**:47-54

[30] Sansavini S, Montevecchi P, Cavicchi C. Experience sui portinesti dellalbicece "Realle de Imola". Riv. Frutticoltura. 1989;**8**(9):59-65

[31] Neumüller M, Muhlberger L, Siegler H. New rootstocks with resistance to *plum pox virus* for *Prunus domestica* and other stone fruit species: The Docera and Dospina rootstock series. Acta Horticulturae. 2013;**985**: 155-165

[32] Weinberger JH. Plums. In: Advances in Fruit Breeding. 1975. pp. 336-347

[33] Dziedzic E, Lech W, Malodobry M, Jaukun A. Studies on fertilization of plum, *in vitro*. Acta Horticulturae. 1998; **478**:113-118

[34] Dziedzic E, Lech W, Maladobry M, Lech M. Observations of ovules of some plum cultivars after pollination, *in vitro* conditions. Acta Horticulturae. 2002;**577**

[35] Lech W, Malodobry M, Lech M, Nowak B. *In vitro* fertilization of plum. ISHS. Acta Horticulturae. 1994;**359**: 269-277

[36] Gercheva PA, Zhivondov A. Embryo rescue of early ripening plum cultivars. Acta Horticulturae. 2002;**577**: 165-168

[37] Scortza R, Callahan AM, Damsteegt V, Levy L, Ravelonandro M. Transferring potyvirus coat protein genes through hybridization of transgenic plants to produce plum pox virus resistant to plums (*Prunus domestica*). Acta Horticulturae. 1998; **472**:421-428

[38] Malinowski T, Cambra M, Capote N, Zawadzka B, Gorris MT, Scorza R, et al. Field trials of plum clones transformed with the plum pox virus coat protein (PPV-CP) gene. Plant Disease. 2006;**90**:1012-1018

[39] Scortza R, Callahan AM, Dardick C. Genetic engineering of plum pox virus resistance: Honey Sweet plum—From concept to product. Plant Cell, Tissue and Organ Culture. 2013;**115**(1):1-12

[40] Ravelonandro M, Malinowski T, Minoiu N, Cambra M. Potential use of transgenic plums resistant to plum pox virus field infection. Acta Horticulturae. 2003;**622**:119-122

[41] Zagrai I, Ravelonandro M, Ioana G, Beatrix F, Scorza R, Zagrai L, et al. Investigation of potential environmental impact of transgenic plums on the diversity of plum pox virus populations. In: XXXVI Annual Meeting Iasi, Romania. Vol. 36/2006. 2006. pp. 1133-1142

[42] Zagrai I, Capote N, Ravelonandro M, Cambra M, Zagrai L, Scorza R. Silencing mechanism of C5 transgenic plums is stable under challenge inoculation of heterologous viruses. Journal of Plant Pathology. 2008;**90**(S1):63-71

[43] Bozhkova V, Zhivondov A. Cultivars commonly used as donors at the breeding for improvement of plum varietal assortment. Plant Sciences. 2004;**41**:51-54

[44] Bozhkova V. Plum genetic resources and breeding. AgroLife Scientific Journal. 2013;**2**(1):83-88

[45] Butac M, Budan S. Evaluation of local plum varieties (*Prunus domestica* L.) from Romanian National Collection. EUCARPIA. XII Fruit Section Symposium. Zaragoza, Spania. Acta Horticulturae. 2009;**814**

[46] Poulos JM. Germplasm collection, evaluation, documentation, and conservation. (A compilation of lecture materials of a training course held in BARI, Joydebpur, Gazipur, Bangladesh, 4–6 May 1992). In: Asian Vegetable Research and Development Center. Shanhua, Tainan, Taiwan. Publication no. 93-398. 1993. p. 95

[47] Thomas TA, Mathur PN. Germplasm evaluation and utilization. In: Paroda RS, Arora RK, editors. Plant Genetic Resources Conservation and Management, Concepts and Approaches. New Delhi, India: IBPGR, Regional Office for South and Southeast Asia; 1991. pp. 149-158

[48] Gass T, Tobutt KR, Zanetto A. Report of working group on *Prunus*; 1996

[49] Maggioni L, Lipman E. Report of working group on Prunus: Sixth and seventh meetings. 2003

[50] Botu M, Botu I, Vicol A, Neagoe A, Vişanu F. Lucrările colocviului național privind gestionarea resurselor genetice din pomicultură. In: Biodiversitatea pomicolă—conservare şi utilizare, la SCDP Vâlcea [Fruit Tree Biodiversity—Conservation and Using at RSFG Valcea]. 2008. pp. 52-55 (in Romanian)

[51] Botu M, Botu I, Achim G, Preda S, Scutelnicu A, Giura S. Conservation of fruit tree genetic resources and their use in the breeding process. The Annals of Valahia University of Targoviste. 2017; **11**(1):66-69

[52] Braniste N, Butac M. Fondul de germoplasmă la speciile pomicole, de arbuşti fructiferi şi căpşun din colecțiile din România [Germplasm Fund of Fruit Trees, Berries and Strawberry from Romanian Collections]. Piteşti: Pământul; 2006. pp. 93-117 (in Romanian)

[53] Butac M, Botu M, Militaru M, Budan S. Romanian germplasm fund and its use in plum breeding program. In: Scientific Papers of RIFG Piteşti

Mărăcineni. Vol. XXVII. Bucharest: Invel Multimedia; 2011. ISSN: 1584-2231

[54] Fideghelli C, Vitellozzi F, Grassi F, Sartori A. Characterization and evaluation of fruit germplasm for a sustainable use. Acta Horticulturae. 2003;**598**:153-160

[55] Vukojevic D, Simic J, Dragisic N, Sevo D, Misimoic M, Zavisic N, et al. Evaluation of the quality of autochthonous plum cultivars in the area of Bosanski Petrovac. In: Third International Scientific Symposium "Agrosym Jahorina 2012". 2012

[56] Sestraş R. *Ameliorarea speciilor horticole* (The Breeding of Fruit Growing Species). Academic Press; 2004. pp. 165-192 (in Romanian)

[57] Butac M, Militaru M, Budan S, Sumedrea D. A survey of the genetic resources used in Romanian plum breeding program. Analele Universității din Craiova. 2010;**XV (XLXI)**:120-126

[58] Eremin GV. Sliva. Kolos: Moskva; 1983

[59] Dan C, Serban C, Sestras A, Militaru M, Morariu P, Sestras R. Consumer perception concerning apple fruit quality, depending on cultivars and hedonic scale of evaluation—A case study. Notulae Scientia Biologicae. 2015; 7(1):140-149

[60] Harker FR, Kupferman EM, Marian AB, Gunson FA, Triggs CM. Eating quality standards for apples based on consumer preferences. Postharvest Biology and Technology. 2008;**50**(1):70-78

[61] Lace B, Lacis G. Evaluation of pear (*Pyrus communis* L.) cultivars in Latvia. Horticultural Science (Prague). 2015; **42**(3):107-113

[62] Seppa L, Railio J, Vehkalahti K, Tahvonen R, Tuorila H. Hedonic responses and individual definitions of an ideal apple as predictors of choice. Journal of Sensory Studies. 2013;**28**: 346-357

[63] Eigenmann C, Kellerhals M. Welche apfel wollen die konsumentinnen und konsumenten? Agrarforshung. 2007;**14**: 6-9

[64] Garcia-Montiel F, Serrano M, Martinez-Romero D, Alburquerque N. Factors influencing fruit set and quality in different sweet cherry cultivars. Spanish Journal of Agricultural Research. 2010;**8**(4):1118-1128

[65] Molnar AM, Ladanyi M, Kovacs S. Evaluation of the production traits and fruit quality of German plum cultivars. Acta Universitatis Agriculturae et Silviculturae Mendelianae Brunensis. 2016;**64**(1):109-114

[66] Budan S, Butac M, Plopa C. Some considerations regarding Romanian plum and cherry germplasm evaluation concerning susceptibility to plum pox virus field infection. In: Lucrari Stiintifice USAMV Iasi, vol I(50) Seria Horticultura. 2007. pp. 397-400

[67] Zagrai I. Diferențierea tulpinilor virusului Plum pox prin TAS-ELISA cu antiseruri monoclonale şi prin IC-RT-PCR cu amorse specifice. In: Lucrari Stiintifice, Ediția II, SCDP Constanta. Bucuresti: Cartea Universitara; 2006. pp. 183-189. (in Romanian)

[68] Hartmann W. The importance of hypersensitivity for breeding plums and prunes resistant to plum pox virus (Sharka). Acta Horticulturae. 2002;**577**: 33-37

[69] Glišić I, Milošević N, Karaklajić-Stajić, Dordević M, Lukić M. Divna—New plum (*Prunus domestica* L.) cultivar developed at Fruit Research Institute, Čačak. Journal of Pomology. 2018; **52**(201):7-13

[70] Milosevič T, Milosevič N. Quantitative analysis of the main biological and fruit quality traits of F1 plum genotypes (*Prunus domestica* L.). Acta Scientiarum Polonorum-Hortorum Cultus. 2011;**10**(2):95-107

[71] Milosevič T, Milosevič N. New plum hybrids resistant to plum pox virus from the Čačak breeding program (Serbia). Comptes Rendus de l'Académie Bulgare des Sciences: Sciences Mathématiques et Naturelles. 2011;**64**(8):1213-1220

[72] Hartmann W. New plum cultivars from Hohenheim. Acta Horticulturae. 1998;**478**:171-174

[73] Hartmann W. Three new Sharka resistant (hypersensitive) plum cultivars from Hohenheim plum breeding programme. Acta Horticulturae. 2019;**1260**:9-13

[74] Jacob H. Twenty-Five years plum breeding in Geisenheim, Germany: Breeding targets and previous realizations. Acta Horticulturae. 2007; **734**:341-346

[75] Jacob H. New plum and mirabelles varieties out of the breeding work and development in Geisenheim. Acta Horticulturae. 2002;**577**:173-176

[76] Zhivondov A, Vitanova I, Ivanova D, Minev I, Stoev A, Blagov A. Major results of plum breeding in Bulgaria. Acta Agriculturae Serbica. 2012;**XVII**(33):3-11

[77] Renaud R, Lafargue B. Ferbleue: A new European plum variety. Acta Horticulturae. 1998;**478**:169-170

[78] Renaud R, Roy M. Selection and evaluation of hybrids issued from intraspecific crosses with *Prunus domestica*. Acta Horticulturae. 1990;**283**: 69-74

[79] Bellini E, Giannelli G, Picadi E. Genetic improvement of European plum: Promising selections from Florence. Riv. Frutticoltura. 1990;**6**: 25-29 (in Italian)

[80] Eremin T, Zaremuk R. Advance of assortment of plums in the South of Russia. Russian Journal of Horticulture. 2014;**1**(1):1-6

[81] Simonov VS, Kulemekov SN. Modern assortment of the plum and ways to increase its adaptivity in Central Region. Proc Sci Works GNBS. 2017; **144**(1):143-150 (in Russian)

[82] Zaremuk RS, Alekhima EM. Breeding of stone fruits (exemplified by sweet cherry and plum) in Southern Russia. Russian Agricultural Sciences. 2013;**39**(2):129-133

[83] Zaremuk RS, Alekhima EM, Bogatyreva SV, Dolya YA. Results of breeding of stone fruit crops under conditions of the South of Russia. Russian Agricultural Sciences. 2017; **43**(4):289-292

[84] Kazlouskaya Z, Matveev VA. New Belarusian plum varieties for modern orchards. In: Eufrin Plum and Prune Working Group Meeting. Research and Breeding Institute of Pomology Holovousy; 2006

[85] Matveev VA, Vasileva MN. Breeding characteristics of plum seedlings of hybrid family 'Narach' × 'Oda' on the complex of economic characters. Plodovodstvo. 2015;**27**:87-92 (in Belarusian)

[86] Janes H, Pae A. Plum breeding at the Polli Horticultural Institute (Estonia). Acta Horticulturae. 1998;**478**:301-304

[87] Kask K, Janes H, Libek A, Arus L, Kikas A, Kaldmae H, et al. New cultivars and future perspectives in professional fruit breeding in Estonia. Agronomy Research. 2010;**8**(Special Issue III): 603-614

[88] Kaufmane E, Gravite I, Trajkovski V. Results of Latvian plum breeding programme. Acta Horticulturae. 2012;**968**:55-60

[89] Kaufmane E, Gravite I, Ikase L. Plum research and growing in Latvia. Proceedings of the Latvian Academy of Sciences Section B. 2019, 2019;**73**(3): 195-206 (720)

[90] Jakubowski T. Breeding of plum cultivars in Poland. Acta Horticulturae. 1998;**478**:151-154

[91] Blažek J, Šecová M. Main characteristics of new plum cultivars bred at Holovousy. Scientia Horticulturae. 2013;**40**(4):149-153

[92] Tehrani G. Seventy five years of plum breeding and pollen compatibility studies in Ontario. Acta Horticulturae. 1990;**283**:95-103

[93] Tehrani G, Lay W. Victory plum. HortScience. 1992;**27**(11):1232-1233

[94] De Jong TM, Castro SJ. Dried plum cultivar development and evaluation. In: Research Report. California Dried Plum Board, UC Davis: Fruit and Nut Research Information; 2015. pp. 4-19

Chapter 2

The North American Plums (*Prunus* Spp.): A Review of the Taxonomic and Phylogenetic Relationships

Dario J. Chavez and José X. Chaparro

Abstract

North America is a center of diversity for *Prunus* species. Tree architecture, chilling requirement, heat requirement, fruit development period, fruit size, fruit texture, disease resistance, and adaptive changes to multiple environmental conditions are a few examples of the traits of which tremendous genetic variability is available in the native plum species. Wild native *Prunus* species constitute an important potential source of genetic diversity for stone fruit breeding and selection. A review of the North American plum taxonomic treatment and phylogenetic studies is described. Various studies have been done with three major groups being identified: Americana series, Chickasaw series, and Beach series.

Keywords: plums, phylogeny, taxonomy, *Prunus*, *Prunocerasus*

1. Introduction

The genus *Prunus* L. belongs to the subfamily Amygdaloideae (=Prunoideae) of the Rosaceae. It has a worldwide distribution with approximately 200 species. Edible species are mostly distributed in the northern hemisphere [1–5]. The genus contains species that are important in the production of fruit, nuts, and lumber. Plums, cherries, almonds, apricots, and peaches are the most commonly known fruit and nuts in this genus. The world's net production of almonds, apricots, cherries, peaches, nectarines, plums, and sloes in 2010 was approximately 40.8 million tons. Peach and nectarine production was the largest in the world with 20.5 million tons. US peach and nectarine production was approximately 1.3 million tons, with a farm gate value of ~683 million dollars [6].

North America is an important center of diversity for plum species adapted (native) to widely divergent climates and soils representing an important potential source of genes for plant breeding. In [7], Layne and Bassi reported high levels of variation in the *Prunus* germplasm for tree size, growth habit, flower size and color, chill hour requirement, fruit size, flesh texture, flesh color, resistance to diseases, and adaptability to diverse climatic and geographic regions. Plums are the stone fruit with the greatest diversity of flavor, aroma, texture, color, form, and size [2, 8].

Stone fruit breeders have used this tremendous genetic variability through interspecific hybridizations (in particular with species in the subgenus *Prunus* or

Prunophora) for the improvement of *Prunus* scion and rootstock cultivars [9]. Among those, native North American plum species have been identified as a source of resistance to blossom blight and brown rot (*Monilinia fructicola*), bacterial spot (*Xanthomonas campestris* pv. *pruni*), bacterial canker (*Pseudomonas syringae* pv. *syringae*), plum leaf scald (*Xylella fastidiosa*), peach tree short life (PTSL), root-knot nematode (*Meloidogyne* spp.), lesion nematode (*Pratylenchus* spp.), clitocybe root rot (*Armillaria tabescens*), and others [9–12].

Resistance to bacterial leaf spot and bacterial canker was identified in a cultivar derived from *P. salicina* Lindl., *P. cerasifera* Ehrh., *P. angustifolia* Marshall, *P. americana* Marshall, and *P. munsoniana* W. Wight & Hedrick. *Prunus hortulana* L.H. Bailey was found resistant to root-knot nematode and has been recommended as a rootstock for European plums. Improved tolerance for PTSL was found in hybrids from *P. americana*, *P. hortulana*, *P. angustifolia*, and/or *P. umbellata* Elliot. Potential uses of the native North American plum species as breeding parents, scions, and/or rootstocks were summarized by [10, 12].

2. Taxonomic treatment

In [8], Waugh described the genus *Prunus* as trees or shrubs, mostly with edible fruit and flowers, white or pink, with spreading petals. Stamens 15–30, distinct, with filiform filaments. Style, terminal; stigma, usually truncate. The fruit has a fleshy exterior, is glabrous, and contains a hard bony pit, which contains the seed.

Inconsistencies in the taxonomy of *Prunus* were recognized by Waugh [8] and Hedrick [2]. Bortiri et al. [1] summarized the classification discrepancies in *Prunus* as follows: (1) four different genera (*Amygdalus*, *Cerasus*, *Prunus*, and *Padus* [13]) and later two (*Amygdalus* and *Prunus*) [14]; (2) five genera (*Amygdalus*, *Persica*, *Prunus*, *Armeniaca*, and *Cerasus* (including *Padus* and *Laurocerasus*)) [15]; (3) *Prunus* as a single genus divided in seven sections (*Amygdalus*, *Armeniaca*, *Prunus*, *Cerasus*, *Laurocerasus*, *Ceraseidos*, and *Amygdalopsis*) [16]; (4) *Prunus* with previous seven sections as subgenera [17]; (5) *Prunus* classified into five subgenera (*Prunophora* (*Prunus*), *Amygdalus*, *Cerasus*, *Padus*, and *Laurocerasus*) and with subgenus *Prunus* divided in three sections (*Euprunus*, *Prunocerasus*, and *Armeniaca*) [3]; and (6) *Prunus* divided into three genera (*Padus*, *Laurocerasus*, and *Prunus*) [18].

Recently, the concept of *Prunus* as single genus has become widely accepted, but subgenera classification is still undistinguished as new phylogenetic relationships within *Prunus* come to light. The USDA-GRIN [19] germplasm collection organizes the genus *Prunus* into subgenus *Amygdalus*, *Cerasus*, *Emplectocladus*, and *Prunus*. Subgenus *Cerasus* was divided into sections *Cerasus* and *Laurocerasus* and subgenus *Prunus* into sections *Armeniaca*, *Microcerasus* (including some plums), *Penarmeniaca*, *Prunocerasus* (the North American plums), and *Prunus*.

Waugh [8] recognized the difficulty in classifying the North American plums and stated "plums grow pretty much as they please, and the botanist has to take them as he finds them." The distribution, cultivation, hybridization, and breeding value of native plums have been extensively studied [2, 4, 5, 8, 20, 21].

Waugh [8] classified the cultivated and indigenous *Prunus* of North America into groups. These groups were clustered into seven series: Americana, Chickasaw, Hortulana, Maritima, Sand Cherry, Choke Cherry, and Black Cherry [22] (**Table 1**). The Americana series included the Americana group (including *P. americana* var. *lanata*) and the Nigra group (*Prunus nigra* Aiton). The Chickasaw series included the Chickasaw and the Sand plum groups. The Hortulana series, categorized as "hybrids," included the Wildgoose group, the Wayland group, and the Miner

	Group	Species	Origin	Cultivation
Cultivated	Domestica plums	*Prunus domestica*	Eastern Europe and west-central Asia	Nova Scotia, central New England, New York, southern Ontario and Michigan, and the Pacific coast states
	Damsons	*Prunus domestica*	Europe	
	Myrobalan plums	*Prunus cerasifera*		Europe and US used as rootstock
	Simon plums	*Prunus simonii*	China	New York, California
	Japanese plums	*Prunus triflora*	China, Japan	Maine, Vermont, Ontario, and southern Iowa
Indigenous	Americana group	*Prunus americana*	USA (Ohio, Texas, northward to Minnesota and Montana)	Prince Edward Island, Manitoba, and Vancouver, to Florida, Louisiana, and Texas
	Nigra group	*Prunus americana nigra*	CAN (Newfoundland west to Rainy and Assiniboine rivers), USA (New England states)	Prince Edward Island, Manitoba, and Vancouver, to Florida, Louisiana, and Texas
	Miner Group	*Prunus hortulana mineri*	USA (standing between *P. americana* and the Wildgoose group)	Not cultivated
	Wayland group	*Prunus rivularis*[z] *Prunus hortulana*[y]	USA (Colorado, Guadalupe, and the Leona)	North of Burlington, Vermont, and Iowa
	Wildgoose group	*Prunus hortulana*	USA (the Mississippi valley)	From Texas to Massachusetts
	Chickasaws	*Prunus angustifolia*	USA (Southern range to Delaware and Kentucky, including southern Atlantic and Gulf states)	Iowa, Vermont, New York, and Massachusetts
	Sand plum	*Prunus angustifolia watsonii*	USA (South and southeast Nebraska and central and western Kansas)	Cultivated by settlers in Kansas and Maryland
	Beach plum	*Prunus maritima*	USA (sea beaches, New Brunswick to Virginia, Georgia, Alabama, and Connecticut)	Not cultivated
	Pacific plum	*Prunus subcordata*	USA (Pacific coast)	Sierra regions of California and southern Oregon

Group	Species	Origin	Cultivation
Oklahoma plum	*Prunus gracilis*	USA (Southern Kansas to Texas and Tennessee)	Not cultivated
Alleghany plum	*Prunus alleghaniensis*	USA (Alleghany mountains in Pennsylvania)	Not cultivated
Southern sloe	*Prunus umbellata*	USA (seashore from South Carolina to Florida and westward to Mississippi, Louisiana, and Arkansas)	Not cultivated
Dwarf cherries	*Prunus pumila* *Prunus pumila besseyi* *Prunus cuneata*	*P. pumila* in USA (coasts of northern states), *P. pumila* besseyi (from Manitoba to Kansas, westward to California and Utah), and *P. cuneata* in USA (New Hampshire to Minnesota and southward to North Carolina)	Nebraska eastward
Choke Cherry	*Prunus virginiana*	CAN (Newfoundland to Manitoba and British Columbia) to USA (Georgia, Texas, and Colorado)	Not cultivated
Black Cherry	*Prunus serotina*	CAN (Quebec) to USA (Kansas and southward, New Mexico, and Mexico)	Not cultivated

[z]*Classified as* Prunus rivularis *but with doubts.*
[y]Prunus hortulana *consider as part of the Wayland and the Wildgoose group.*

Table 1.
Cultivated and indigenous plums in North America by group, area of origin, and cultivation [8].

group. The Maritima series the Beach plum group, the Southern sloe group [including *P. umbellata* Elliot var. *injuncunda* (Small) Sarg.], the Oklahoma plum group, and *P. glandulosa* Thunb. (ungrouped). The Sand Cherry series were equivalent to the Dwarf cherries group. The Choke Cherry and the Black Cherry series conserved their name as groups [8, 22] (**Table 1**).

Wight [5] separated the genus *Prunus* in plums, cherries, and dwarf cherries. Waugh's [8, 22] taxonomic treatment included cherries as part of plums. Wight's [5] groups/series were Americana, Subcordata, Hortulana, Angustifolia, Maritima, and Gracilis. The Angustifolia group agreed with Waugh's [22] Chickasaw series. Waugh [22] did not include *P. mexicana S. Watson* (Americana group), *P. munsoniana* (Angustifolia group), *P. subcordata* Benth. (Subcordata group), *P. alleghaniensis* Porter (Maritima group), and *P. umbellata* (Maritima group), as part of his groups/series.

3. *Prunus* phylogenetic studies

Phylogeny and systematics in the genus *Prunus* was reported by [23]. They employed isozymes to study the phylogenetic relationships in *Prunus*. Section *Prunocerasus* was found to be polyphyletic, with a clade formed by *P. americana*, *P. munsoniana*, *P. hortulana*, *P. subcordata*, and *P. angustifolia*, and a clade formed by *P. maritima* Marshall and *P. umbellata*.

Chloroplast DNA is an alternative source of genetic variation and is maternally inherited in *Prunus*. Chloroplast DNA is highly conserved and in relative abundance in the cell as compared with the nuclear DNA. Kaneko et al. [24] and Uematsu et al. [25] used cpDNA to classify cherries, apricots, and wild and cultivated peaches in Japan. In [26], Badenes and Parfitt reported a phylogeny similar to Mowrey and Werner [23]. All the *Prunus* species were grouped as in conventional subgenus classifications [3]. *Prunus persica* L.-*P. dulcis* (Mill.) D.A. Webb, *P. domestica* L.-*P. salicina* Lindl., and *P. cerasus* L.-*P. fruticosa* Pall were monophyletic.

Lee and Wen's [27] phylogenetic analysis of the genus *Prunus* using ITS sequences recognized two major groups: the *Amygdalus-Prunus* group, and the *Cerasus-Laurocerasus-Padus* group. The results were not congruent with Rehder's [3] taxonomic treatment.

In Bortiri et al. [1] the phylogeny and systematics of *Prunus* based on ITS and chloroplast *trnL-trnF* spacer DNA sequences identified two major clades: subgenera *Padus-Laurocerasus-Cerasus* and subgenera *Prunus-Amygdalus-Emplectocladus-Cerasus* (sect. *Microcerasus*)-sect. *Penarmeniaca* (similar to Mowrey and Werner [23], Lee and Wen [27], and Bortiri et al. [1]). Their results indicated that plums of northeastern North America were closely related and that *P. mexicana* belonged to a sister clade.

Bortiri et al. [28] used the nuclear gene *s6pdh*, which encodes $NADP^+$-dependent sorbitol-6-phosphate dehydrogenase, to assess the lack of support for deep nodes in the clade subgenera *Prunus-Amygdalus-Emplectocladus* (as reported in previous data). The phylogenies based on ITS, cpDNA *trnL-trnF*, and *s6pdh* sequences were compared and combined. Phylogenetic analysis of the combined data supported two major clades: subgenera *Cerasus-Laurocerasus-Padus* and subgenera *Amygdalus-Emplectocladus-Prunus*. Section *Microcerasus* (subgenera *Cerasus*) was reported nested within subgenus *Prunus*.

Prunus subg. *Prunus* sect. *Prunocerasus* was reported to be monophyletic by Shaw and Small [29]. The phylogenetic analysis was based on seven cpDNA regions: *rp*S16, *rp*L16, *trn*L, *trn*G, *trn*L-*trn*F, *trn*S-*trn*G, and *trn*H-*psb*A. Three clades were strongly supported in sect. *Prunocerasus*: the "American Clade," the "Chickasaw Clade," and the "Beach Clade" (names based on Waugh's (1901) classification). The American clade included *P. americana* Marshall var. *americana* Sudw., *P. americana* Marshall var. *lanata*, *P. mexicana*, *P. rivularis* Scheele, *P. hortulana*, and *P. umbellata* var. *injucunda*; the Chickasaw clade included *P. angustifolia*, *P. munsoniana*, *P. gracilis* Engelm. & A. Gray, *P. nigra*, *P. umbellata* Elliot var. *umbellata*, *P. alleghaniensis* Porter var. *alleghaniensis*, and *P. alleghaniensis* Porter var. *davisii* (W. Wight) Sarg.; and the Beach clade included *P. geniculata* Harper, *P. maritima* Marshall var. *maritima*, and *P. maritima* Marshall var. *gravesii* (Small) G.J. Anderson.

Similarly, a survey of cpDNA haplotypes available within section *Prunocerasus* was reported by Shaw and Small [30]. The cpDNA *rpL*16 region was sequenced for 207 accession representatives of 17 North American plums, including *P. texana* D. Dietr. (as described before). More than one of the three primary cpDNA haplotypes was found in many of the taxa.

Bortiri et al. [31] studied the evolution of vegetative and morphological characters of 37 species of *Prunus* and other genera of Rosaceae. Morphological characters were combined with ITS, *trn*L-*trn*F, and *trn*S-*trn*G data from previous studies [1, 28]. The addition of the morphological data with *trn*S-*trn*G supported some nodes that were found in ITS and *trn*L-*trn*F studies. Three clades were reported: "Clade A" with subgenera *Padus* and *Laurocerasus*; "Clade B" with subgenera *Amygdalus*, *Emplectocladus*, and *Prunus*; and "Clade C" with subgenera *Cerasus*. "Clade B" was characterized by the production of three axillary buds. *Padus* and *Laurocerasus* were not supported as monophyletic (high homoplasy).

Genetic diversity within *Prunus cerasifera* (cherry plum) was studied using morphological characters, cytometry, cpDNA, and SSR markers [32]. Morphological characters showed differences between clones. Analysis of cpDNA reported 15 haplotypes clustered in 3 groups. Considerable diversity among accessions was reported based on these studies.

Paper[z]	Kaneko et al. [24]	Mowrey and Werner [23]
Phylogenetic analysis	Molecular	Molecular
Analytical methods	Phenetics—percent differential restriction fragments and Engel's genetic distance	Phenetics—principal components
Metrics (analysis)	cpDNA using *BamHI*, *Hind*III, and *Sma*I	Isozyme
Taxa (no.)/ subgenus (sect.) genus	11 species/3 subgenus: *Cerasus*, *Padus*, *Armeniaca* [3]/genus *Prunus*	34 species/4 subgenus: *Prunus* (sect.: *Prunus*, *Prunocerasus*, *Armeniaca*), *Amygdalus*, *Cerasus* (sect.: *Sargentiella*, *Microcalymma*, *Magniculpula*, *Phyllomahaleb*), and *Lithocerasus* (sect.: *Microcerasus*, *Armeniacocerasus*) [35]
Outgroups		
Trees (no.)	2	2 (average 30 principal components)
Characters or bp (no.)		
Informative characters (no.)		
Indels (no.)		
Substitutions (no.)		
Inversions (no.)		
PIC		
Percent variability		
Phylogeny in classification		Support for subgenus *Prunus.* Subgenus *Lithocerasus* was identified as an artificial grouping of species
Notes		*Lithocerasus* formed part of *Cerasus* in Rehder's [3] classification

Paper	Badenes and Parfitt [26]	Lee and Wen [27]
Phylogenetic analysis	Molecular	Molecular
Analytical methods	MP	MP, NJ, ML
Metrics (analysis)	cpDNA cutting with 21 3.2 kb and 10 2.1 kb endonucleases	ITS nuclear ribosomal DNA
Taxa (no.)/ subgenus (sect.)	9 species/5 subgenus: *Prunus*, *Amygdalus*, and *Cerasus.*	40 species (represented by 52 accessions)/5 subgenus: *Prunus* (sect.: *Prunus*, *Prunocerasus, Armeniaca*), *Amygdalus*, *Cerasus* (sect.: *Microcerasus*, *Pseudocerasus*, *Mahaleb*, *Phyllomahaleb*), *Padus*, and *Laurocerasus* [3]
Outgroups	*Fragaria vesca*	*Exochorda giraldii*, *Maddenia hypoleuca*, *Oemleria cerasiformis*, *Prinsepia sinensis*, *Prinsepia uniflora*, *Lyonothamnus floribundus*

Paper	Badenes and Parfitt [26]	Lee and Wen [27]
Trees (no.)	10	MP = 15,000 MPT (L = 630, CI = 0.632, RC = 0.510). Consensus tree 16,383 MPTs (L = 630, CI = 0.632, RI = 0.808). ML tree log likelihood = −3641.3155
Characters (no.)	23	662 bp aligned (ITS1 = 223–242 bp, 5.8 s = 154 bp, and ITS2 = 201–219 bp)
Informative characters (no.)		218 bp aligned (ITS1 = 114 bp, 5.8 s = 12 bp, and ITS2 = 92 bp)
Indels (no.)		29 indels (>3 bp) aligned (ITS1 = 13 bp, ITS2 = 16 bp)
Substitutions (no.)		
Inversions (no.)		
PIC		218 bp aligned (ITS1 = 114 bp, 5.8 s = 12 bp, ITS2 = 92 bp) (not including indels)
Percent variability		32.9% aligned (ITS1 = 47.1%, 5.8 s = 7.79%, ITS2 = 42.0%)
Phylogeny in classification	Support for subgenus *Prunus*, *Cerasus*, and *Amygdalus*. Relative small number of taxa used in the study. Subgenus *Cerasus* suggested to be more extensively evolved than either *Prunus* or *Amygdalus*	Genus *Prunus* was monophyletic. Support for *Maddenia* nested within genus *Prunus*. Within genus *Prunus*, two major groups were recognizable: *Amygdalus-Prunus* group and *Cerasus-Laurocerasus-Padus* group
Notes		Number of parsimony informative characters included outgroups. The % variability cannot be directly compared with studies that excluded the outgroups for the number of PICs
Paper	**Bortiri et al. [1]**	**Bortiri et al. [28]**
Phylogenetic analysis	Molecular	Molecular
Analytical methods	MP	MP, ML
Metrics (analysis)	ITS nuclear ribosomal DNA and chloroplast *trn*L-*trn*F spacer DNA	Nuclear gene sorbitol 6-phosphate dehydrogenase (*s6pdh*) and data from previous study ITS and *trn*L-*trn*F [1]
Taxa (no.)/ subgenus (sect.)	48 species/5 subgenus: *Prunus* (sect.: *Prunus, Prunocerasus, Armeniaca*), *Amygdalus*, *Cerasus* (sect.: *Microcerasus, Pseudocerasus, Mahaleb, Phyllomahaleb*), *Padus*, and *Laurocerasus* [3]	22 species (representing all the major clades found in previous study)/5 subgenus: *Prunus* (sect.: *Prunus, Prunocerasus, Armeniaca*), *Amygdalus*, *Cerasus* (sect.: *Microcerasus, Pseudocerasus, Mahaleb, Phyllomahaleb*), *Padus*, and *Laurocerasus* [3]
Outgroups	*Exochorda racemosa*, *Oemleria cerasiformis*, *Prinsepia sinensis*, *Physocarpus capitatus*, *Sorbaria sorbifolia*, and *Spiraea cantoniensis*	*Exochorda racemosa*, *Oemleria cerasiformis*, *Sorbaria sorbifolia, Spiraea cantoniensi, Holodiscus microphyllus*, *Chamaebatiaria millefolium*, *Kageneckia oblonga*, *Vauquelinia californica*, *Gillenia stipulata, Pyrus caucasica*, *Sorbus* sp., *Amelanchier alnifolia*, *Aruncus dioicus*, *Neilla sinensis*, and *Spiraea betulifolia*

Paper	Bortiri et al. [1]	Bortiri et al. [28]
Trees (no.)	*trn*L-*trnF* sequence—MP = 76 MPT (L = 187, CI = 0.733, RI = 0.834). ITS sequence—MP = stopped at 30000 MPT (L = 678, CI = 0.567, RI = 0.714). Combined data set—consensus tree 8318 MPT (L = 876, CI = 0.695, RI = 0.727)	*s6pdh* sequence—MP = 273 MPT (L = 1198, CI = 0.58, RI = 0.81). s6pdh sequence—ML tree log likelihood = −7720.96. For combined data set—MP = 9 MPT (L = 1592, CI = 0.58, RI = 0.61). For combined data set—ML tree log likelihood = −12056.56
Characters (no.)	*trn*L-*trn*F = 563 bp, ITS = 759 bp	*s6pdh* = 1387 bp. Combined data set = 2760 bp (*s6pdh, trn*L-*trn*F, and ITS)
Informative characters (no.)	*trn*L-*trn*F = 26 bp (excluding outgroups), ITS = 76 bp (excluding outgroups = among *Prunus* species)	*s6pdh* = 234 bp (excluding outgroups = among *Prunus* species). Combined data set = 226 bp (*s6pdh* = 148, *trn*L-*trn*F = 18, and ITS = 60)
Indels (no.)	*trn*L-*trn*F = 9 indels (>2 bp), ITS = 2 indels (>2 bp)	
Substitutions (no.)		
Inversions (no.)		
PIC	*trn*L-*trn*F = 26 bp (excluding outgroups), ITS = 76 bp (excluding outgroups = among *Prunus* species) (not including indels)	*s6pdh* = 234 bp (excluding outgroups = among *Prunus* species). Combined data set = 226 bp (s6pdh = 148, *trn*L-*trn*F = 18, and ITS = 60)
Percent variability	*trn*L-*trn*F = 4.62%, ITS = 10.01%	*s6pdh* = 16.87%. For combined data set = 8.18% (*s6pdh* = 10.67%, *trn*L-*trn*F = 3.19%, and ITS = 7.9% = calculated with characters from Bortiri et al. [1])
Phylogeny in classification	Genus *Prunus* was monophyletic. *Exochorda*, *Oemleria*, and *Prinsepia* were not supported as sister groups with *Prunus*. Genus *Prunus* was divided in two clades: subgenera *Amygdalus-Prunus-Cerasus* (sect. *Microcerasus*)-*Emplectocladus* group and subgenera *Cerasus-Laurocerasus-Padus* group. Subgenus *Prunus* sect. *Prunus* was monophyletic	Genus *Prunus* was monophyletic. In the combined data set, the genus *Prunus* was formed by two groups: subgenera *Cerasus-Laurocerasus-Padus* and subgenera *Amygdalus-Emplectocladus-Prunus-Cerasus* (sect. *Microcerasus*)
Notes	First time that *P. fasciculata* (sect. *Emplectocladus*) was used in a study	Includes *P. fasciculata* sect. *Emplectocladus*

Paper	Shaw and Small [29]
Phylogenetic analysis	Molecular
Analytical methods	MP, BI
Metrics (analysis)	Seven noncoding chloroplast DNA regions: *trnL*^{UAA}, *rpS*16, *rpL*16, and *trnG*^{UUC} introns; *trnS*^{GCU}-*trn*G^{UUC}; *trnL*^{UUA}-*trnF*^{GAA}; and *trnH*^{GUG}-*psbA* intergeneric spacers
Taxa (no.)/ subgenus (sect.)	43 species/5 subgenus: *Prunus* [sect.: *Prunus*, *Prunocerasus* (17 taxa), *Armeniaca*], *Amygdalus*, *Cerasus* (sect.: *Microcerasus*, *Pseudocerasus*, *Mahaleb*, *Phyllomahaleb*), *Padus*, and *Laurocerasus* [3]
Outgroups	*Physocarpus opulifolius*
Trees (no.)	Combined data set—MP = 25,171 MPT (L = 422, CI = 0.92, RI = 0.94)

Paper	Shaw and Small [29]
Characters or bp (no.)	*Prunocerasus* analysis introns: *trnL*^{UAA} = 522 bp, *rpS16* = 683 bp, *rpL16* = 996 bp, and *trnG*^{UUC} = 711 bp. Intergeneric spacers: *trnS*^{GCU}-*trnG*^{UUC} = 703 bp, *trnL*^{UUA}-*trnF*^{GAA} = 397 bp, and *trnH*^{GUG}-*psbA* = 363 bp. Combined data = 4375 bp. *Prunus* analysis *trnH*^{GUG}-*psbA* = 516 bp, *rpL16* = 1105 bp, *trnS*^{GCU}-*trnG*^{UUC} = 903 bp, and *trnG*^{UUC} = 746 bp. Combined data = 3270 bp
Informative characters (no.)	
Indels (no.)	*Prunocerasus* analysis introns: *trnL*^{UAA} = 0 bp, *rpS16* = 2 bp, *rpL16* = 7 bp, and *trnG*^{UUC} = 0 bp. Intergeneric spacers: *trnS*^{GCU}-*trnG*^{UUC} = 2 bp, *trnL*^{UUA}-*trnF*^{GAA} = 0 bp, and *trnH*^{GUG}-*psbA* = 3 bp. Combined data = 14 bp. *Prunus* analysis *trnH*^{GUG}-*psbA* = 13 bp, *rpL16* = 10 bp, *trnS*^{GCU}-*trnG*^{UUC} = 14 bp, *trnG*^{UUC} = 4 bp. Combined data = 41 bp
Substitutions (no.)	*Prunocerasus* analysis introns: *trnL*^{UAA} = 1 bp, *rpS16* = 4 bp, *rpL16* = 6 bp, and *trnG*^{UUC} = 4 bp. Intergeneric spacers: *trnS*^{GCU}-*trnG*^{UUC} = 4 bp, *trnL*^{UUA}-*trnF*^{GAA} = 3 bp, and *trnH*^{GUG}-*psbA* = 1 bp. Combined data = 23 bp. *Prunus* analysis *trnH*^{GUG}-*psbA* = 11 bp, *rpL16* = 21 bp, *trnS*^{GCU}-*trnG*^{UUC} = 28 bp, and *trnG*^{UUC} = 32 bp. Combined data = 92 bp
Inversions (no.)	*Prunocerasus* analysis introns: *trnL*^{UAA} = 0 bp, *rpS16* = 0 bp, *rpL16* = 0 bp, and *trnG*^{UUC} = 0 bp. Intergeneric spacers: *trnS*^{GCU}-*trnG*^{UUC} = 0 bp, *trnL*^{UUA}-*trnF*^{GAA} = 0 bp, and *trnH*^{GUG}-*psbA* = 0 bp. Combined data = 0 bp. *Prunus* analysis *trnH*^{GUG}-*psbA* = 0 bp, *rpL16* = 0 bp, *trnS*^{GCU}-*trnG*^{UUC} = 1 bp, and *trnG*^{UUC} = 0 bp. Combined data = 1 bp
PIC	*Prunocerasus* analysis introns: *trnL*^{UAA} = 1 bp, *rpS16* = 6 bp, *rpL16* = 13 bp, and *trnG*^{UUC} = 4 bp. Intergeneric spacers: *trnS*^{GCU}-*trnG*^{UUC} = 6 bp, *trnL*^{UUA}-*trnF*^{GAA} = 3 bp, and *trnH*^{GUG}-*psbA* = 4 bp. Combined data = 37 bp. *Prunus* analysis *trnH*^{GUG}-*psbA* = 24 bp, *rpL16* = 31 bp, *trnS*^{GCU}-*trnG*^{UUC} = 43 bp, and *trnG*^{UUC} = 36 bp. Combined data = 134 bp
Percent variability	*Prunocerasus* analysis introns: *trnL*^{UAA} = 0.19%, *rpS16* = 0.88%, *rpL16* = 1.31%, and *trnG*^{UUC} = 0.56%. Intergeneric spacers: *trnS*^{GCU}-*trnG*^{UUC} = 0.85%, *trnL*^{UUA}-*trnF*^{GAA} = 0.76%, and *trnH*^{GUG}-*psbA* = 1.10%. Combined data = 37 bp. *Prunus* analysis *trnH*^{GUG}-*psbA* = 4.65%, *rpL16* = 2.80%, *trnS*^{GCU}-*trnG*^{UUC} = 4.76%, and *trnG*^{UUC} = 4.80%. Combined data = 4.09%.
Phylogeny in classification	Genus *Prunus* was monophyletic. Subgenus *Prunus* sect. *Prunocerasus* and sect. *Prunus* were monophyletic. The genus *Prunus* was formed by two groups: subgenera *Laurocerasus-Padus* and subgenera *Amygdalus-Emplectocladus-Prunus-Cerasus* (sect. *Microcerasus*). *Prunus texana* and *P. subcordata* were included in sect. *Prunocerasus*. Within sect. *Prunocerasus* three groups were identified: the American, the Chickasaw, and the Beach clades
Notes	*Prunus texana* was first used in this study. *Prunus texana* and *P. fasciculata* were not recognized by Waugh [8], Wight [5], and Rehder [3]

Paper	Rohrer et al. [36]	Shaw and Small [30]	Katayama and Uematsu [37]
Phylogenetic analysis	Molecular	Molecular	Molecular
Analytical methods	UPGMA	MP	UPGMA
Metrics (analysis)	Fifteen microsatellites primer pairs	*rpL16* intron	CpDNA analysis based on five restriction enzymes (*Sal*I, *Xho*I, *Bam*HI, *Sac*I, and *Pst*I) by RFLP
Taxa (no.)/ subgenus (sect.)/genus	18 species/subgenus *Prunus* sect. *Prunocerasus* (13 and 3 undetermined hybrids), subgenus *Prunus* (*P. cerasifera*), and	A total of 207 accessions = 18 species (subgenus *Prunus* sect. *Prunocerasus*)	A total of 18 accessions = 14 *Prunus* species and 1 interspecific hybrid

Paper	Rohrer et al. [36]	Shaw and Small [30]	Katayama and Uematsu [37]
	subgenus *Armeniaca* (*P. armeniaca*).		
Outgroups			*Pyrus ussuriensis* var. *hondoensis*
Trees (no.)		Strict consensus = 3 MPT (L = 34, CI = 0.97, RI = 0.99)	Strict consensus = 8 MPT (L = 68, CI = 0.93, RI = 0.64)
Characters or bp (no.)	A total of 186 putative alleles with a mean value of 12.4 per locus	*rpL*16 intron = 797 bp	
Informative characters (no.)		*rpL*16 intron = 23 bp	
Indels (no.)			
Substitutions (no.)			
Inversions (no.)			
PIC		*rpL*16 intron = 23 bp	
Percent variability		*rpL*16 intron = 2.88%	
Phylogeny in classification	No clear phylogenetic relationships were determined. The microsatellites are evolving too rapidly in North American plums to be truly useful at resolving species relationships	Twenty-two unique haplotypes were identified in sect. *Prunocerasus*. Ten different haplotypes were associated with the American clade, two haplotypes with the Beach clade, and seven haplotypes with the Chickasaw clade. Additionally, one Texana haplotype, one Subcordata haplotype, and one peculiar *Umbellata* haplotype	Eleven genome types. The UPGMA tree consisted of two major groups: genome types A-I (subgenus *Amygdalus*, *Prunus*, and *Cerasus* sect. *Microcerasus*) and other with genomes J-K (subgenus *Laurocerasus* and *Padus*).
Notes	The congeneric relationship of plums to peach and cherry allowed the successful use of these primers in section *Prunocerasus*. Microsatellites are evolving too rapidly to be truly useful at resolving species phylogeny	The common practice of choosing one specimen to represent a taxon can be misleading in closely related groups. Choosing different genotypes could have resulted in a different result than previous studies	The 9.1 kb region between *psb*A and *atp*A genes would be useful tool to study the cpDNA evolution in *Prunus*

Paper	Bortiri et al. [31]	Wen et al. [38]
Phylogenetic analysis	Morphology and molecular	Molecular
Analytical methods	MP, ML, and BI	MP and BI

Paper	Bortiri et al. [31]	Wen et al. [38]
Metrics (analysis)	ITS nuclear ribosomal gene, *trn*L-*trn*F spacer, *trn*S-*trn*G spacer, *trn*G intron, and 25 morphological characters.	Chloroplast *ndh*F region and ITS nuclear ribosomal gene.
Taxa (no.)/ subgenus (sect.)/genus	37 species/5 subgenus: *Prunus* (sect.: *Prunus*, *Prunocerasus*, *Armeniaca*), *Amygdalus*, *Cerasus* (sect.: *Microcerasus*, *Pseudocerasus*, *Mahaleb*, *Phyllomahaleb*), *Padus*, and *Laurocerasus* [3]	A total of 59 (*ndh*F) or 51 (ITS) accessions of *Prunus*/5 subgenus: *Prunus* (sect.: *Prunus, Prunocerasus, Armeniaca*), *Amygdalus, Cerasus* (sect.: *Microcerasus, Pseudocerasus, Mahaleb*, *Phyllomahaleb*), *Padus*, and *Laurocerasus* [3]. In addition, *Madenia hypoleuca* and the *Pygeum* group
Outgroups	*Oemleria cerasiformis*, *Sorbaria sorbifolia, Spiraea cantoniensi, Gillenia stipulata, Lyonothamnus floribundus*, *Maddenia hypoleuca*, *Physocarpus capitatus*, *Physocarpus opulifolius*, and *Rhodotypos scandens*	*Oemleria cerasiformis*, *Prinsepia uniflora*, *Physocarpus monogynus*, *Lyonothamnus floribundus*, and *Holodiscus discolor*
Trees (no.)	Morphological data set—MP = 50,000 MPT (L = 110, CI = 0.36, RI = 0.73). Molecular data results from Bortiri et al. [1] and Bortiri et al. [28]. Combined data set—MP = 20 MPT (L = 1741, CI = 0.49, RI = 0.65). Combined data set—ML tree log likelihood = 12499.63	*ndh*F sequence—MP = 196,200 MPT (L = 815, CI = 0.71, COI = 056, RI = 0.86). ITS sequence—MP = 49,200 MPT (L = 791, CI = 0.56, COI = 0.45, RI = 0.70)
Characters or bp (no.)	Combined data set = 771 bp	
Informative characters (no.)	ITS = 178 bp, *trn*L-*trn*F = 50 bp, and *trn*S-*trn*G = 142 bp	
Indels (no.)	Combined data set = 3	
Substitutions (no.)		
Inversions (no.)		
PIC	ITS = 178 bp, *trn*L-*trn*F = 50 bp, and *trn*S-*trn*G = 142 bp	
Percent variability		
Phylogeny in classification	Three clades were reported: "Clade A" with subgenera *Padus* and *Laurocerasus*; "Clade B" with subgenera *Amygdalus*, *Emplectocladus*, and *Prunus*; and "Clade C" with subgenera *Cerasus*. "Clade B" was characterized by the production of three axillary buds. *Padus* and *Laurocerasus* were not supported as monophyletic (highly homoplasy)	Both data set identified genus *Prunus* as a monophyletic group. Both data sets were incongruent at the species level in *Prunus*. The *ndh*F data supported two major groups: subgenera *Laurocerasus* (including *Pygeum*) and *Padus*, and subgenera *Amygdalus*, *Cerasus*, and *Prunus*. The ITS data supported a clade composed of subgenera *Amygdalus*, *Prunus,* and *Cerasus* sect. *Microcerasus*, and the paraphyletic clade of *Padus* and *Laurocerasus*
Paper	**Depypere et al. [33]**	**Chavez et al. [39]**
Phylogenetic analysis	Morphology and molecular	Molecular
Analytical methods	UPGMA, PCo, and BI	MP and ML

Paper	Depypere et al. [33]	Chavez et al. [39]
Metrics (analysis)	Leaf and endocarp morphometrics and AFLP primers	SSRs (41), cpDNA (seven regions), nuclear genes (33 vernalization response genes, 16 tree architecture, and 3 isozymes), and ITS
Taxa (no.)/ subgenus (sect.)/genus	A total of 82 accessions/5 species: *P. cerasifera*, *P. domestica*, *P. insititia*, *P. spinosa*, and *P.* × *fruticans*,	A total of 8 species: *P. americana*, *P. angustifolia*, *P. hortulana*, *P. mexicana*, *P. munsoniana*, *P. geniculata*, *P. maritima*, *P. umbellata*
Outgroups		*P. fasciculata*, *P. persica*, and *P. pumila*
Trees (no.)		cpDNA sequences—MP = 13 MPT (L = 623, CI = 0.92, RI = 0.81, RC = 0.74) – ML = −lnL = 5414.74. Nuclear genes – MP = 1 MPT (L = 2535, CI = 0.88, RI = 0.88, RC = 0.78) – ML = −lnL = 41509.34. Combined nuclear + cpDNA + ITS – MP = 2 MPT (L = 2732, CI = 0.88, RI = 0.88, RC = 0.77) – ML = −lnL = 48496.34.
Characters or bp (no.)		Combined data set = 27,623 bp
Informative characters (no.)		1594
Indels (no.)		
Substitutions (no.)		
Inversions (no.)		
PIC		
Percent variability		
Phylogeny in classification	PCoA and AFLP of three distinct clusters. A first cluster consists of all *P. cerasifera* samples and the sole *P. cocomilia*. A second cluster includes all individuals of *P. domestica* and *P. insititia*. A third cluster comprises all *P. spinosa* and *P.* × *fruticans* samples	The American and the Chickasaw clades were identified. An outgroup clade was comprised by *P. persica* and *P. fasciculata*
Notes	Low number of *Prunus* species for sampling	Identified multiple gene regions that provided the greatest number of characters, variability, and improved phylogenetic signal at the species level in *Prunus* section *Prunocerasus*

[z]PIC = total indels + nucleotide substitutions + inversions. Percent variability = PIC/characters or bp. PIC = potentially informative character.

Table 2.
Summary of Prunus *phylogenetic studies.*

Endocarp and leaf morphometrics combined with AFLP markers were used to study the morphological and genetic variation of five European members of section *Prunus*: *P. cerasifera*, *P. cocomilia* Ten., *P. domestica*, *P. insititia* L., *P. spinosa* L., and *P.* × *fruticans* [33]. Three clusters were reported: a first cluster *P. cerasifera*-*P. cocomilia*, a second *P. domestica*-*P. insititia*, and a third *P. spinosa* and *P.* × *fruticans*.

Phylogenetic analysis based on four single-copy cpDNA regions (*atp*B-*rbc*L, *mat*K, *rpl*16, and *trn*L-*trn*F) of Eurasian plums, *Prunus* section *Prunus*, confirmed this section to be monophyletic. Four well supported clades were reported: "Clade A" with *P. salicina*, *P. sogdiana*, and *P. ussuriensis*; "Clade B" with *P. cocomilia*; "Clade C" with *P. brigantina*, *P. ramburii*, and *P. spinosa*; and "Clade D" with subclade D1 *P. domestica*-*P. insititia*-*P. divaricata*-*P. ursine* and subclade D2 *P. cerasifera* [34].

Chavez et al. [39] identified genomic regions that provided the greatest number of characters and variability and improved the phylogenetic signal at the low level in *Prunus* section *Prunocerasus* relationships. The American and the Chickasaw clades were identified. An outgroup clade was comprised by *P. persica* and *P. fasciculata*. The results reported were similar to those reported by Mowrey and Werner [23].

Previous studies demonstrated the value of morphology, cytometry, nuclear DNA, and cpDNA as data for phylogenetic studies in *Prunus*. Most of the previous phylogenetic research used Mason's [21] and Rehder's [3] taxonomic classification. A complete summary of *Prunus* phylogenetic research is summarized in **Table 2**.

4. Final remark

The subgenus *Prunus* section *Prunocerasus* (the North American plums) constitutes important genetic resources (gene pool) of unique traits such as tree architecture, chilling requirement, heat requirement, fruit development period, fruit size, fruit texture, disease and insect resistance, and adaptive changes to multiple environmental conditions, among others. These species could be used in the breeding of improved stone fruit cultivars in the future. The summary of the taxonomic and phylogenetic relationships presented in this chapter provides a base to understand the species relationships. In addition, it will help for the conservation and maintenance of a broader germplasm base within *Prunus*.

Author details

Dario J. Chavez[1]* and José X. Chaparro[2]

1 University of Georgia, USA

2 University of Florida, USA

*Address all correspondence to: dchavez@uga.edu

References

[1] Bortiri E, Oh S-H, Jiang J, Baggett S, Granger A, Weeks C, et al. Phylogeny and systematic of *Prunus* (Rosaceae) as determined by sequence analysis of ITS and the chloroplast *trnL-trnF* spacer DNA. Systematic Botany. 2001;**26**: 797-807

[2] Hedrick UP. The plums of New York. N.Y. Dept. Agr. 18th Ann. Rpt. v. 3, pt. 2. 1911

[3] Rehder A. Manual of Cultivated Trees and Shrubs Hardy in North America. 2nd ed. New York: Macmillan; 1940

[4] Sargent CS. Manual of the Trees of North America. Cambridge: The Riverside Press; 1905

[5] Wight WF. Native American species of *Prunus*. Bulletin of the U.S. Department of Agriculture. 1915;**179**: 1-75

[6] FAOSTAT Data. 2010. Available from: http://faostat3.fao.org/home/

[7] Layne DR, Bassi D. The peach. In: Botany, Production and Uses. Cambridge, MA: CABI International; 2008. pp. 615

[8] Waugh FA. Plums and Plum Culture. New York, USA: Orange Judd; 1901

[9] Layne RE, Sherman WB. Interspecific hybridization of Prunus. HortScience. 1986;**21**:48-51

[10] Beckman TG, Okie WR. Native North American plum species potential for variety and rootstock development. Acta Horticulturae. 1994;**359**:41-48

[11] Beckman TG, Okie WR, Nyczepir AP, Pusey PL, Reilly CC. Relative susceptibility of peach and plum germplasm to Armillaria root rot. HortScience. 1998;**33**:1062-1065

[12] Okie WR, Weinberger JH. Plums. In: Janick J, Moore JN, editors. Fruit Breeding, Tree and Tropical Fruits. Vol. 1. New York: John Wiley and Sons, Inc.; 1996

[13] Linnaeus. Hortus Cliffortianus. Amsterdam, Holland: Published by the Author; 1737

[14] Linnaeus. Genera Plantarum. 5th ed. Stockholm: Impensis Laurentti Salvii; 1754

[15] De Candolle AP. Prodromus Systematis Naturalis Regni Vegetabilis. Vol. 2. Paris: Treuttel et Würtz; 1825

[16] Bentham G, Hooker JD. Genera Plantarum. Vol. 1. London: Lovell Reed & Co.; 1865

[17] Koehne E. Deutsche Dendrology. Sttugart: Verlag von Ferdinand Enke; 1893

[18] Hutchinson J. The Genera of the Flowering Plants, Dicotyledones. Vol. 1. Oxford: Clarendon Press; 1964

[19] USDA. Plants-Profile Database (PLANTS) [Online Database]. Available from: http://plants.usda.gov/index.html [Accessed: 07 April 2010]

[20] Britton NL, Shafer JA. North American Trees Being Descriptions and Illustrations of the Trees Growing Independently of Cultivation in North America, North Mexico and the West Indies. New York: New York, USA; 1908

[21] Mason SC. The pubescent-fruited species of Prunus of the southwestern states. Journal of Agricultural Research. 1913;**1**:147-179

[22] Waugh FA. The pollination of plums. Vermont Agricultural Experimentation Station, Annual Report. 1899;**12**:189-209

[23] Mowrey BD, Werner DJ. Phylogenetic relationships among species of *Prunus* as inferred by isozyme markers. Theoretical and Applied Genetics. 1990;**80**:129-133

[24] Kaneko T, Terachi T, Tsunewaki K. Studies on the origin of crops species by restriction endonuclease analysis of organellar DNA. II. Restriction analysis of ctDNA of 11 Prunus species. Japanese Journal of Genetics. 1986;**61**:157-168

[25] Uematsu C, Sasakuma T, Ogihara Y. Phylogenetic relationships in the stone-fruit group of Prunus as revealed by restriction fragment analysis of chloroplast DNA. Japanese Journal of Genetics. 1991;**66**:59-69

[26] Badenes ML, Parfitt DE. Phylogenetic relationships of cultivated Prunus species from an analysis of chloroplast DNA variation. Theoretical and Applied Genetics. 1995;**90**: 1035-1041

[27] Lee S, Wen J. A phylogenetic analysis of *Prunus* and the *Amygdaloideae* (Rosaceae) using ITS sequences of nuclear ribosomal DNA. American Journal of Botany. 2001;**88**: 150-160

[28] Bortiri E, Oh S-H, Gao F-Y, Potter D. The phylogenetic utility of nucleotide sequences of sorbitol 6-phosphate dehydrogenase in Prunus (Rosaceae). American Journal of Botany. 2002;**89**:1697-1708

[29] Shaw J, Small RL. Addressing the "hardest puzzle in the American pomology:" Phylogeny of *Prunus* sect. Prunocerasus (Rosaceae) based on seven noncoding chloroplast DNA regions. American Journal of Botany. 2004;**91**:985-996

[30] Shaw J, Small RL. Chloroplast DNA phylogeny and phylogeography of the north American plums (*Prunus* subgenus *Prunus* section *Prunocerasus*, Rosaceae). American Journal of Botany. 2005;**92**:2011-2030

[31] Bortiri E, Vanden HB, Potter D. Phylogenetic analysis of morphology in Prunus reveals extensive homoplasy. Plant Systematics and Evolution. 2006; **259**:53-71

[32] Horvath A, Christmann H, Laigret F. Genetic diversity and relationships among *Prunus cerasifera* (cherry plum) clones. Botany. 2008;**86**: 1311-1318

[33] Depyrere L, Chaerle P, Breyne P, Mijnsbrugge KV, Goetghebeur P. A combined morphometric and AFLP based diversity study challenges the taxonomy of the European members of the complex *Prunus* L. section *Prunus*. Plant Systematics and Evolution. 2009; **279**:219-231

[34] Reales A, Sargent DJ, Tobutt KR, Rivera D. Phylogenetics of Eurasian plums, *Prunus* L. section *Prunus* (Rosaceae), according to coding and non-coding chloroplast DNA sequences. Tree Genetics and Genomes. 2010;**6**: 37-45

[35] Krüssmann G. Manual of Cultivated Broad-Leaved Trees and Shrubs. Vol. 3. Portland: Timber Press; 1986

[36] Rohrer JR, Ahmad R, Southwick SM, Potter D. Microsatellite analysis of relationships among North American plums (*Prunus* sect. *Prunocerasus*, Rosaceae). Plant Systematics and Evolution. 2004;**244**: 69-75

[37] Katayama H, Uematsu C. Structural analysis of chloroplast DNA in *Prunus* (Rosaceae): Evolution, genetic diversity and unequal mutations. Theoretical and Applied Genetics. 2005;**111**:1430-1439

[38] Wen J, Berggren ST, Lee C-H, Ickert-Bond S, Yi T-S, Yoo K-O, et al. Phylogenetic inferences in *Prunus*

(Rosaceae) using chloroplast *ndhF* and nuclear ribosomal ITS sequences. Journal of Systematics and Evolution. 2008;**46**:322-332

[39] Chavez DJ, Beckman TG, Chaparro JX. Identifying the North American plum species phylogenetic signal using nuclear, mitochondrial, and chloroplast DNA markers. Journal of the American Society for Horticultural Science. 2016;**141**:623-644

Section 2

Fruit Tree Physiology

Chapter 3

Response of Growth Inhibitor Paclobutrazol in Fruit Crops

Naira Ashraf and Moieza Ashraf

Abstract

Paclobutrazol (PBZ; IUPAC name: (2RS, 3RS)-1-(4-chlorophenyl)-4, 4-dimethyl-2-(1H-1, 2, 4-triazol-1-yl) pentan-3-ol) is a triazol derivative and an antagonist of gibberellins. It has been shown to inhibit shoot growth in various perennial fruit trees. Paclobutrazol application reduced the number of shoots, transforming trees into a more desirable, spur-type growth habit as the vegetative sink was reduced. This compound induces an early and intense flowering, diminishing vegetative growth and reducing the extension of buds, allowing for ripening and the initiation of apical buds inflorescence. Besides, it also increases fruit set, the years following application as a carryover effect. An increase in return bloom is a common response to paclobutrazol treatment and has been reported for various fruit crops. Paclobutrazol is widely used to advance harvest maturity in various fruit crops and it improves fruit quality in terms of accelerated colour development, delayed and synchronized fruit maturation and control of preharvest fruit drop. It is known to improve fruit physical and fruit chemical characteristics. Fruit calcium is increased for 2–3 years due to carry over effect. It helps in the maintenance of better fruit quality during storage and influences nutrient uptake in various fruit crops including stone fruits. It has been characterized as an environmentally stable compound in soil and water environments with a long half-life under both aerobic and anaerobic conditions.

Keywords: fruit set, paclobutrazol, flowering, shoot growth, fruit quality

1. Introduction

One of the most important elements in fruit orchard management is growth control. Excessive vigour reduces light penetration, yield, fruit quality and an increase in cost of pruning and pest control [1]. On the other hand, many cultivars may set very large number of fruits with unacceptably smaller size and often serious reduction in return bloom and fruit set may occur in the following year leading to biennial bearing. About 20–80% of the fruit which initially sets drop-off from the tree during various stages of development. Thus, various plant growth regulators are used for the control of vegetative growth, flowering of young trees, thinning of flowers and fruits, delaying fruit abscission, regulation of fruit ripening and improvement of fruit production and quality in bearing trees [2]. Various chemicals are used in horticulturally advanced countries to reduce the amount of pruning. Among the various growth control chemicals tested, paclobutrazol (PB) is one of the most successful and widely used in fruit trees that can retard tree growth. Besides this, it also increases fruit set the years following application as a carryover effect. Paclobutrazol treatments have also shown to increase Delicious fruit firmness at harvest, [3]. Paclobutrazol, a gibberellins inhibitor, has been effectively used in reducing canopy volume and increasing flower intensity in peach [4], plum [5], almond [6], grapes [7] and mango [8]. Paclobutrazol is effective not only in flower induction but also in early and off season flower induction in mango [9, 10]. Paclobutrazol application in McIntosh apple trees shortly after full bloom affected fruit quality characteristics with respect to accelerated colour development, delayed and synchronised fruit maturation, control of preharvest fruit drop and maintenance of better fruit quality during storage [11]. Sebastian et al. [12] also reported that the foliar application of plant growth regulators improves the yield and quality of fruit crops. However, the action of plant growth regulators (PGRs) is highly specific to plant species, cultivar and stage of development, and strongly dependent on its rate of application and environmental conditions [13].

2. Nature of paclobutrazol

The plant growth regulator, [(2RS, 3RS)-1-(4-chlorophenyl)-4,4-dimethyl-2-(1,2,4-triazol-1-yl)-pentan-3-ol] (paclobutrazol; PP333) with chemical formula $C_{15}H_{20}ClN_3O$, is a triazole derivative and has been shown to inhibit shoot growth on apple trees [14]. The ancymidol blocks the oxidative steps with high specificity leading from ent-kaurene to ent-kaurenoic acid in the pathway of GA biosynthesis. The same oxidative steps are thought to be inhibited by the active triazol derivatives. Paclobutrazol has been reported to inhibit GA biosynthesis in plants by inhibiting kaurene oxidase, a Cyt P-450 oxidase, thus blocking the oxidation of kaurene to kaurenoic acid [15]. The inhibitory activity of paclobutrazol can be reversed by GA [14]. Besides reducing gibberellins level, PBZ increases cytokinin contents, root activity and C:N ratio, whereas its influence on nutrient uptake lacks consistency [16]. Paclobutrazol was also shown to shift assimilate partitioning from leaves to roots, increase carbohydrates in all parts of apple seedlings, increase chlorophyll content, soluble protein and mineral element concentration in leaf tissue, increase root respiration and reduce water use [17]. Browning et al. [18] investigated the effect of PBZ on the translocation of endogenous IAA (indol-3-acetic acid) in Doyenne du Comice pear cultivar and found that PBZ caused a slower movement of IAA in shoot tips. The usual application of paclobutrazol has been either by foliar spray or soil drenching. With foliar spray applications, absorption through mature leaves was limited and PBZ may be taken up through stem absorption [19] or from excess dripping onto the soil. Elfving and Proctor [20] have reported that protecting the soil from foliar drip reduced the PBZ-induced inhibition of extension growth in apples. When applied to the soil, a continuous supply of PBZ taken up by the roots is translocated acropetally via the xylem, thus maintaining the concentration of PBZ above the threshold required for the inhibition of gibberellins biosynthesis [21], although phloem translocation has also been reported [22].

Paclobutrazol (PBZ), a non-polar broad spectrum growth regulator, has been characterised as an environmentally stable compound in soil and water environments with a long half-life under both aerobic and anaerobic conditions. Moreover, PBZ is unlikely to volatilise to any significant extent owing to a low estimated vapour pressure (1.9×10^{-6} Pa). Paclobutrazol has been registered in 1985 (cultar, ICI Americas, Goldsboro, NC), however it has now been permitted for use on food crops in Australia, New Zealand, South Africa, India, Philippines,

Vietnam, Canada, USA (California), Finland, Hungry, Greece, Cyprus, Denmark and Netherlands [23]. In India PBZ has been registered as a plant growth regulator under the Section 9(3) of Insecticides Act, 1968 in November 2009 by Central Insecticides Board & Registration Committee [24] and is available in the market with various trade names.

3. Tree growth and vigour

Paclobutrazol when applied during early summer has been observed as effective suppressant of stem growth in sweet cherry [25]. Young [26] reported that paclobutrazol when sprayed on 'Redhaven' cultivar of peach reduced terminal growth and advanced leaf fall. Webster and his associates [27] observed that application of 1.6 g a.i. tree^{-1} paclobutrazol to cherry trees followed by 0.8 g a. i. paclobutrazol in next year inhibited extension growth of young trees on either colt or FB_{22} rootstocks. Gaash [28] stated that 1000–4000 ppm of paclobutrazol on 'Canino' apricot cultivar decreased the lateral shoot length. Foliar spray of 1000 ppm paclobutrazol to sweet cherry trees suppressed shoot growth and delay in fruit colour [29]. Blanco [30] stated that paclobutrazol decreased the shoot development of 'Crimson Gold' nectarines. Kaska et al. [31] observed that application of paclobutrazol decreased shoot elongation in cherries when applied on the vegetative and reproductive parts. Shoot extension growth was reduced by 57% at stone hardening and by 47.6% at harvest following soil and collar drench of cultar (2.0 g a.i. PP333/tree) applied in autumn in peach cv. Flordaprince [32]. Kuden et al. [33] observed that 250 ppm of paclobutrazol decreased the shoot growth by about 34.1–42.4% in apricot. Leaf chlorophyll of almonds was increased with the application of 50 and 100 ppm paclobutrazol [34]. Arzani and Roosta [35] reported that paclobutrazol significantly reduced vegetative growth in apricot. They also reported that the total pruning dry weight, shoot growth and trunk cross sectional area (TCSA) of treated trees were lower than those of the control. Pant and Kumar [36] reported that the application of different concentrations of paclobutrazol and chloromequat at 250, 500, 1000 and 5000 ppm to 'Red Delicious' apple decreased the extension growth and leaf area of trees. Wani [37] applied paclobutrazol through soil to the basin of tree on trunk diameter basis. The investigation was carried out under two independent experiments. In one experiment, treatments were applied for two years consecutively and in another during first year only in order to assess carryover effect during second year. Application of paclobutrazol decreased yearly increment of tree trunk cross-sectional area, tree spread, volume and internodal length in sweet cherry. Gupta and Bist [38] noticed that soil application of paclobutrazol @ 10 ml/lit on high density pear plantation effectively controlled the excessive and vigorous growth. They also observed that the vegetative growth was inhibited by paclobutrazol. Asin et al. [39] observed that paclobutrazol resulted in shortest shoot length in pear. Sharma and Joolka [40] recorded reduced extension growth, plant height and plant spread with paclobutrazol in nonpareil almond plants. Abdollahi et al. [41] reported that paclobutrazol reduced vegetative growth by reducing both fresh and dry weights of shoots and the leaf area was also significantly decreased in strawberry cv. Selva. Mir et al. [42] reported that paclobutrazol significantly retarded the shoot growth, shoot diameter and trunk cross-sectional of 'Roundel' apricot trees growing under low density planting system. Ashraf et al. [43, 44] revealed that apple trees cv. Red Delicious treated with paclobutrazol @ 750 ppm and summer pruning resulted in minimum vegetative growth and vigour in terms of tree height, trunk diameter, annual shoot extension growth, tree spread and leaf area in comparison to control and other treatments. Reduction in growth is attributed to paclobutrazol, which is a triazol that inhibits gibberellin biosynthesis especially three steps in the oxidation of the GA precursors ent-kaurene to ent-kaurenoic acid [15, 45].

4. Fruit set and yield

Webster et al. [27] reported that application of paclobutrazol @ 1.6 g a.i. tree $^{-1}$ in 'Early Rivers' sweet cherry doubled the floral buds per unit shoot length in 1982 and trebled the number in 1983. Both the number of floral buds per fruiting spur and the number of flowers per floral bud increased by 26.8 and 5.6%, respectively. However, Webster and Quinlan [46] applied paclobutrazol to European plum trees and noticed reduction in yield and this effect was partially alleviated by sprays of 75 mg GA_3 + 10 mg 2, 4, 5-TP at fruit set. Gaash [28] stated that the application of 1000–4000 ppm of paclobutrazol on 'Canino' apricot cultivar increased the yield. Increased number of vegetative and floral buds in 'Flavorcrest' peach was noticed by Martin et al. [47] by the application of 1.32 g a.i. paclobutrazol. Blanco [30] stated that paclobutrazol increased the average fruit size and yield of 'Crimson gold' nectarines. Stan et al. [48] revealed that foliar and soil application of paclobutrazol enhanced the flower bud formation and fruit set in high density planting of sweet cherry, peach and plum. In avocado, paclobutrazol enhanced the fruit set by increasing the partitioning of dry matter to fruits [49]. Kaska et al. [31] observed that application of paclobutrazol decreased flower bud formation in cherries when applied on the vegetative and reproductive parts. Kuden et al. [33] observed that 250 ppm of paclobutrazol increased the number of fruit buds in 'Canino' and 'Precoce de Colomer' apricot cultivars. Jindal and Chandel [50] reported that the application of paclobutrazol increased fruit set in 'Santa Rosa' plum and they further observed that maximum fruit set was observed by the application of 20 ppm TRIA followed by 500 ppm paclobutrazol. Arzani et al. [51] reported that paclobutrazol application advanced flowering of 5-year-old vigorous 'Sundrop' apricot trees by 2–4 days and also increased the fruit set, final fruit number, crop density and yield efficiency. Pant and Ratan [36] reported maximum number of fruit spurs, bloom per branch and yield with cultar @ 1000 ppm in 'Red Delicious' apple. Increased flowering by paclobutrazol application is because of the fact that paclobutrazol acts as inhibitor of gibberellin biosynthesis which changes the sink-source relationship by reallocating carbohydrate to other organs. Davenport [52] reported that more gibberellins were exported from apple fruit to spurs in biennial bearing cultivars than in regular flowering cultivars, concluding that endogenous gibberellins have inhibitory effect on flowering. Increased flowering by application of paclobutrazol was also reported by Wani [37] in sweet cherry and Pant and Ratan [36] in 'Red Delicious' apple. Application of paclobutrazol initiates flowering in fruit plants by the decrease of gibberellins levels and increase of auxins and cytokinins levels in shoot tip [53]. They also observed that flowering date was advanced slightly with application of paclobutrazol. Similar results were recorded by [8, 54–57] in mango. Carreno et al. [58] reported that the application of paclobutrazol before blooming increased fruit set in grapes. The higher fruit set by paclobutrazol application may be attributed to increased partitioning of dry matter to fruits [49]. Abdel Rahim et al. [54] observed that paclobutrazol application advanced off season flowering of regular bearing mango cultivars, Baladi Abu Zaid (26.7%) and Baladi Burai (30.7%) by almost 60 and 70 days, respectively, as compared to the control. They also reported that the flowering percentage in the paclobutrazol treated trees were 50%, and 100% at 60, and 90 days, respectively for all tested cultivars. Similar results on the positive effects of paclobutrazol on mango flowering were reported in many tropical and subtropical regions of the world [16, 55]. Ashraf et al. [43, 44] reported that the highest yield of 101.0 kg tree^{-1} was obtained in treatment 750 ppm paclobutrazol +2 summer prunings in comparison to control in Red Delicious apples. Paclobutrazol (80 ml/tree) produced earlier flowers (125.79 days) with respect to panicle emergence in mango cv. Alphonso compared to control (165.04 days) [59]. Paclobutrazol is effective in enhancing the yield of several horticultural crops as it inhibits gibberellic acid (GA) biosynthesis which changes the sink-source relationship by reallocating carbohydrate to other organs [53]. Huang et al. [60] reported that soil application of paclobutrazol twice during the year 1989 and 1991 in spring at 1.5 g a.i and 0.75 g a.i., respectively increased the yield efficiency twice than control in apple cultivar 'Aki Fuji' (with average fruit yield of 26.25 kg tree^{-1} compared with 13.95 kg tree^{-1}) which might be due to increased respiration or activation of enzymes or growth promoting substances. Kumar et al. [61] reported that the application of paclobutrazol at 1.0 g in October enhances yield and quality in mango.

5. Return bloom

George and Nissen [62] observed that the return bloom was increased in peach in the subsequent season of paclobutrazol application which may result in a large number of small fruit if a large percentage of flowers set. Asin et al. [63] observed that paclobutrazol and root pruning increased return bloom and yield in 'Blanquilla' pear. Asin et al. [39] reported that foliar application of paclobutrazol resulted in highest return bloom in 'Blanquilla' pear. Wani et al. [64] observed that the soil application of paclobutrazol increased the return bloom significantly in Red Delicious apples. Bill [65] noticed that paclobutrazol application increased the average number of flowers per shoot compared to the control in the 2010–2011 seasons. Flower numbers also increased linearly with an increase in paclobutrazol application rate and noticed that paclobutrazol application increased the return bloom. Application of paclobutrazol initiates flowering in fruit plants by the decrease of gibberellins levels and increase of auxins and cytokinins levels in shoot tip. An increase in return bloom is a common response to paclobutrazol treatment and its application has a carryover effect on return bloom as well which has been reported for various fruit crops, such as peach [66], apple [67] and mango [10].

6. Fruit physical characteristics

Paclobutrazol (PBZ), one such GA inhibitor, is widely used to advance harvest maturity in various fruit crops including mango [68], peach [62] and persimmon [69]. Delayed fruit maturation and increased fruit weight was found in peach by the application of paclobutrazol [70]. Webster et al. [27] reported that application of 1.8 g a.i paclobutrazol per tree of 'Early Rivers' cherry applied at full bloom stage increased the fruit weight. Looney and Mckeller [29] observed that application of 1.15 g paclobutrazol per tree increased weight of individual cherry fruits in the year of application and for the following 3 years. Martin et al. [47] reported that application of paclobutrazol @ 0.5, 0.75, 1.0 and 2.0 kg ha^{-1} in 'Flavorcrest' peach increased the size of fruits significantly than control. Blanco [30] stated that paclobutrazol increased the average fruit size of 'Crimson gold' nectarines. Blanco [71] noticed that 2 g a.i. paclobutrazol dissolved in 1 litre of water and pouring the solution around the trunk of 'Crimson Gold' nectarine tree increased the weight of fruit though not significantly. Jindal and Chandel [50] applied paclobutrazol in 'Santa Rosa' plum at 125, 250 and 500 ppm once at full bloom and again at pit hardening stage and reported maximum fruit weight of 24.33 g and fruit volume of 21.6 cc in fruits treated with 500 ppm of paclobutrazol. In persimmons, soil drench application of paclobutrazol accelerated ripening by 2–3 weeks [69]. The increase in fruit length and breadth was due to the reason that application of paclobutrazol reduced vegetative growth (sinks) which in turn, increased the partitioning of nutrients and dry matter towards fruits and thereby, increased the fruit size and weight [49]. Greene [72] reported that foliar application of paclobutrazol to Delicious apples produced fruits with higher flesh and less bitter pit, cork spot and senescence breakdown. Wani [37] reported that fruit acidity, vitamin C, percentage of bruised fruits, incidence of pitting and fruit cracking were reduced by the application of paclobutrazol in sweet cherry. Also, the organoleptic rating, total soluble solids, reducing sugars, total sugars and anthocyanin were increased. Pant and Ratan [36] studied the influence of different concentrations of paclobutrazol and chloromequat at 250, 500, 1000 and 5000 ppm on quality of apple cv. Red Delicious and observed that fruit weight and firmness was increased with both growth retardants. In contrast, in strawberries paclobutrazol application rate had no significant effect on fruit firmness [73]. Carreno et al. [58] found that grape berry size increased linearly with an increase in paclobutrazol application rate. Ashraf et al. [43, 44] observed that treatment 750 ppm paclobutrazol +2 summer prunings resulted in significantly improved fruit size (53.15 cm), weight (188.19 g), volume (188.12 cm^3), colour change (3.40 score), firmness (11.98 kg cm^{-2}), organoleptic rating in terms of taste (3.14 score), texture (3.24 score), flavour (3.12 score) and total soluble solids (14.47°B) whereas acidity (0.23%) was reduced in comparison to control and other treatments during both the years in apple cv. Red Delicious. The improvement in organoleptic rating of fruits may be attributed to the fact that more metabolites were translocated to the fruits in treated trees with paclobutrazol.

7. Fruit colour

Application of paclobutrazol @ 0.33, 0.50, 0.66 and 1.32 g a.i. as soil application to 'Flavorcrest' peach hastened the fruit colour than control [47]. Looney and Mckeller [29] reported that paclobutrazol either sprayed once with 1000 ppm or twice with 500 ppm concentration to 'Lambert' cherry displayed less red colour as indicated by juice anthocyanin concentration or by visual rating of skin colour and 500 mg l^{-1} paclobutrazol within 5 weeks after full bloom to 'Mclntosh' apples gave high percentage of fruit with acceptable red colour at harvest [74]. Santa Rosa plum trees treated with 500 ppm paclobutrazol once at full bloom and repeated at pit hardening stage recorded maximum anthocyanin content (0.299 OD units) which was significantly higher than control [50]. Wani [37] observed that fruit colour was enhanced by the application of paclobutrazol in sweet cherry. Continuous application of paclobutrazol significantly reduces vegetative growth characters of the trees, thereby exposing fruits to direct sunlight which significantly increased red colouration of the fruits. The soil application of paclobutrazol in 'Red Delicious' apple fruits increased the fruit anthocyanin by the increasing dose of paclobutrazol [64].

8. Fruit chemical characteristics

Jindal and Chandel [50] observed that application of 500 ppm paclobutrazol to Santa Rosa plum in two successive years increased total sugars significantly from 5.46 to 6.71% in first year and 6.18 to 7.15% in next year. Similarly, reducing sugars also increased to 4.76 and 5.25% respectively, which was significantly higher than control and recorded an average of 3.94 and 4.50% reducing sugars. Also, least acid contents of 2.14 and 3.35% were observed than control which recorded average acid content of 2.45 and 3.13%, respectively. Wani [37] reported that fruit acidity, vitamin C, percentage of bruised fruits, incidence of pitting and fruit cracking were reduced by the application of paclobutrazol in sweet cherry. Also, the organoleptic rating, total soluble solids, reducing sugars, total sugars and anthocyanin were increased. Wani et al. [64] observed that the soil application of paclobutrazol decreased the acidity and ascorbic acid of 'Red Delicious' apple fruits. Also, the fruit total soluble solids, organoleptic rating and fruit calcium was increased by the increasing dose of paclobutrazol. Similar findings were noticed by Sarker et al. [75] in mango. Paclobutrazol application reduced the number of shoots, transforming trees into a more desirable, spur type growth habit and as the vegetative sink was reduced, transport of nutrients including calcium towards fruits was enhanced [72]. Higher uptake of Ca and its relocation to fruits could be attributed to significantly reduced rate of leaf transpiration, thus could favour the supply of Ca towards the fruit [76]. Fruit calcium is increased for 2–3 years due to carry over effect. Andres et al. [77] observed that the acidity content of fruits diminished as a result of the ripening process and the mango fruits treated with paclobutrazol and KNO_3 showed the lowest values for acidity. Ashraf et al. [78] reported that with increase in paclobutrazol concentration and pruning levels, an increase in TSS, TSS/acid ratio, anthocyanin, sugars, fruit calcium and improvement in fruit grade was observed with decrease in fruit acidity in Red Delicious apples. This increased total soluble solids was due to increased sucrose, starch and sugar levels due to reduced vegetative growth and thus the absence of other potentially competitive actively growing sinks which resulted in more nutrient partitioning to fruits [54]. The increased rate of photosynthesis led by more light penetration into the interior tree canopy, increased the soluble solids in fruits harvested from pruned trees. These findings are in conformity with the findings of Kumar et al. [61] in mango.

9. Fruit storage behaviour

Wolstenholme et al. [49] reported that the application of paclobutrazol increased the partitioning of nutrients and dry matter towards fruits and thereby, increased the fruit size and weight. This increase in weight reduced the physiological loss in weight of fruits during storage period. The differences in storage performance may be due to ethylene production, responsible for the changes in texture and firmness and fruit softening [79]. Elfving et al. [80] observed that the fruits of McIntosh apple treated with diaminozide and paclobutrazol and stored for 24 weeks were firmer and displayed less core browning than untreated ones. Elfving et al. [74] reported that foliar application of 500 mg l^{-1} paclobutrazol in McIntosh apples reduced flesh firmness loss and reduced post storage ethylene production in one season. Later applications at 5 and 9 weeks after full bloom affected stem cavity browning with increase in brown core. This may be attributed to reduction in ethylene evolution during storage which induced delay in respiratory climacteric after harvest and storage thereby, the loss in firmness was decreased [81]. The fruit flesh was firm due to retardation in ripening. Several physiological disorders and diseases of apple fruit during storage are related to the calcium content of fruit [82]. Calcium deficiency results in economic losses in fruit. It helps in regulation of metabolism in apple fruit and adequate concentrations maintain fruit flesh firmness and minimise the incidence of physiological disorders like water core, bitter pit and internal breakdown [83]. The increase in calcium generally delays the ripening of the fruit and maintains their quality during prolonged storage. Fruit calcium is increased for 2–3 years due to carry over effect by paclobutrazol application. This increase in fruit calcium reduced bitter pit, cork spot, senescent breakdown so, spoilage of fruits was reduced which in turn enhanced storage life of fruits [17]. Our findings are in conformity with the findings of Wani [37] in sweet cherry and Wani et al. [64] in 'Red Delicious' apple. Paclobutrazol application reduced the number of shoots, transforming trees into a more desirable, spur type growth habit and as the vegetative sink was reduced, transport of nutrients including calcium towards fruits was enhanced.

10. Nutrient uptake

Paclobutrazol application influences the leaf nutrient status of various temperate fruit crops:

Nitrogen: Paclobutrazol application reduced foliar N concentration in Nemaguard [84], Flordaprince [85], Flordaprince and Flordagold peach cultivars [86] and Red Spur Delicious and Vance Delicious apples [87]. However, Sharma and Joolka [40] recorded reduced leaf N content with paclobutrazol in nonpareil almond plants.

Phosphorus: Paclobutrazol application reduced foliar P concentration in Nemaguard [84], Flordaprince [85], Flordaprince and Flordagold peach cultivars [86] and Red Spur Delicious and Vance Delicious apples [87]. Increased foliar P concentration in apple plants treated with PP333 has been reported by Curry [88]. However, Sharma and Joolka [40] recorded reduced leaf P content with paclobutrazol in nonpareil almond plants.

Potassium: Paclobutrazol application reduced foliar K contents in Nemaguard [84], Flordaprince peach [85], stone fruits [89], Red Spur Delicious and Vance Delicious [87]. Contrary to this, Swietlik and Miller [90] observed increase in K uptake with the addition of 0.2 ppm PP333 to a nutrient solution in which 11 month old apple seedlings were grown. However, Sharma and Joolka [40] recorded reduced leaf K content with paclobutrazol in nonpareil almond plants.

Calcium: Increased concentration of foliar Ca with paclobutrazol application was observed in Nemaguard [84], Flordaprince [85], Flordaprince and Flordagold peach cultivars [86] and Red Spur Delicious and Vance Delicious apples [87]. Similar observations regarding the increase in foliar Ca concentrations in various apple cultivars were made by Bonomo et al. [91]. Swietlik and Miller [92] further reported that Ca content in Golden Delicious increased in proportion to the increasing doses of PP333. Sharma and Joolka [40] also recorded increased leaf Ca content with paclobutrazol in nonpareil almond plants.

Magnesium: Foliar Mg content has been reported to increase with paclobutrazol application in Nemaguard [84], Flordaprince [85], Flordaprince and Flordagold peach cultivars [86], Red Spur Delicious and Vance Delicious apples [87] and apple plants [91]. But Curry [88] found reduced levels of foliar Mg in apple plants treated with PP333. However, Sharma and Joolka [40] also recorded increased leaf Mg content with paclobutrazol in nonpareil almond plants.

11. Conclusion

Paclobutrazol (PP333) has been effectively used to manipulate tree vigour in several perennial fruit crops. Paclobutrazol application reduces the amount of pruning which requires skilled labour and is time consuming and costly.

Paclobutrazol is effective not only in flower induction but also in early and off season flower induction which thereby maintains regularity and synchronisation in flowering. Paclobutrazol application affected fruit yield, quality characteristics and helps in maintenance of better fruit quality during storage. It influences the nutrient uptake in various fruit crops. Paclobutrazol was also shown to shift assimilate partitioning from leaves to roots, increase carbohydrates in all parts of fruit seedlings which enhances cold hardiness during winter periods. It increases chlorophyll content, soluble protein and mineral element concentration in leaf tissue which result in compact darker leaves. It also increases root respiration and reduce water use hence such trees are suitable under drought conditions.

Author details

Naira Ashraf[1*] and Moieza Ashraf[2]

1 Division of Fruit Science, Sher-e-Kashmir University of Agricultural Sciences and Technology of Kashmir, Srinagar, India

2 Division of Environmental Science, Sher-e-Kashmir University of Agricultural Sciences and Technology of Kashmir, Srinagar, India

*Address all correspondence to: naira.ashraf@gmail.com

References

[1] Miller SS. Root pruning and trunk scoring have a limited effect on young bearing apple trees. HortScience. 1995;**30**:981-984

[2] Negi ND, Sharma N. Growth, flowering and cropping response of Flemish beauty pear to bloom spray of GA and BA. Acta Horticulturae. 2005;**696**:295-298

[3] Greene DW. Effect of paclobutrazol and analogous on growth, yield, fruit quality, and storage potential of delicious apples. Journal of the American Society for Horticultural Science. 1986;**11**:328-332

[4] Allan P, George AP, Nissen RJ, RasmussenTS,Morley-BunkerMJ.Effects of paclobutrazol on phenological cycling of low-chill 'Flordaprince' peach in subtropical Australia. Scientia Horticulturae. 1993;**53**:73-84

[5] Olivier OJ, Jacobs G, Strydom DK. Effect of a foliar application of paclobutrazol in autumn on the reproductive development of 'Songold' plum. South African Journal of Plant and Soil. 1990;7:92-95

[6] Koukourikou-Petridou MA. Paclobutrazol affects the extension growth and the levels of endogenous IAA of almond seedlings. Plant Growth Regulation. 1996;**18**:187-190

[7] Christov C, Tsvetkov I, Kovachev V. Use of paclobutrazol to control vegetative growth and improve fruiting efficiency of grapevines (*Vitis vinifera* L.). Bulgarian Journal of Plant Pysiology. 1995;**21**:64-71

[8] Nartvaranant P, Subhadrabandhu S, Tongumpai P. Practical aspects in producing off season mango in Thailand. Acta Horticulturae. 2000;(509):661-668

[9] Burondkar MM, Rajan S, Upreti KK, Reddy YTN, Singh VK, Sabale SN, et al. Advancing Alphonso mango harvest season in lateritic rockysoils of Konkan region through manipulation in time of paclobutrazol application. Journal of Applied Horticulture. 2013;**15**:178-182

[10] Nafees M, Faqeer M, Ahmad S, Khan MA, Jamil M, Aslam MN. Paclobutrazol soil drench suppresses vegetative growth, reduces malformation, and increases production in mango. International Journal of Fruit Science. 2010;**10**:431-440

[11] Edgerton LJ, Blanpied GD. Interaction of succinic acid 2, 2-dimethylhydrazide, 2-choloroethylphosphoric acid and auxins on maturity, quality and abscission of apples. Journal of the American Society for Horticultural Science. 1970;**95**:664-666

[12] Sebastian K, Arya MS, Reshma UR, Anaswara SJ, Thampi SS. Impact of plant growth regulators on fruit production karishma. International Journal of Current Microbiology and Applied Sciences. 2019;**8**(2):800-814

[13] Hoffmann G. Use of plant growth regulators in arable crops: Survey and outlook. In: Karssen CM, Van LC, Vreugdenhil D, editors. Proceedings of 14th International Conerence on Plant Growth Substances. London, UK: Kluwer Academic Publications; 1992. pp. 798-808

[14] Steffens GL, Wang SY, Faust M, Byun JK. Growth, carbohydrate and mineral element status of shoot and spur leaves and fruit of 'Spartan' apple trees treated with paclobutrazol. Journal of the American Society for Horticultural Science. 1985;**110**(6):850-855

[15] Dalziel J, Lawrence DK. Biochemical and biological effects of kaurene oxidase inhibitors such as paclobutrazol. Monograph of British Plant Growth Regulators Group. 1984;**11**:43-57

[16] Kishore K, Singh HS, Kurian RM. Paclobutrazol use in perennial fruit crops and its residual effects: A review. Indian Journal of Agricultural Sciences. 2015;**85**(7):863-872

[17] Wang CY, Steffens GL. Postharvest responses of 'Spartan' apples to preharvest paclobutrazol treatment. HortScience. 1987;**22**:276-278

[18] Browning G, Kuden A, Blanke P. Site of (2RS, 3RS) paclobutrazol promotion of axillary flower initiation in pear cv. Doyenne du Comice. Journal of Horticultural Science. 1992;**67**(1):121-128

[19] Barrett JE, Bartuska CA. PP333 effects on stem elongation dependent on site of application. HortScience. 1988;**17**:737-738

[20] Elfving DC, Proctor JTA. Long-term effects of paclobutrazol (cultar) on apple tree shoot growth, cropping and fruit-leaf relation. Acta Horticulturae. 1986;**179**:473-480

[21] Davies TD, Steffens GL, Sankhla N. Triazole plant growth regulators. Horticultural Reviews. 1988;**10**:63-96

[22] Witchard M. Paclobutrazol is phloem mobile in castor oil plant (*Ricinus communis* L.). Journal of Plant Growth Regulation. 1997;**16**:215-217

[23] Davis T, Curry E. Chemical regulation of vegetative growth. Critical Reviews in Plant Sciences. 1991;**10**:204-206

[24] Kegley SE, Hill BR, Orme S, Choi AH. PAN Pesticide Database. North America, San Francisco: Pesticide Action Network; 2010

[25] Quinlan JD, Webster AD. Effect of the growth retardant PP333 on the growth of plums and cherries. In: Abstract XXist International Horticultural Congress Hamburgh Germany Federal Republic; International Society for Horticulture Science. Vol. 1. 1982. p. 1071

[26] Young PS. Peach growth response from PP_{333} (Paclobutrazol). In: Proceedings of Tenth Annual Meeting Plant Growth Regulator Society of America; East Lansing, USA. PGR Society of America. 1983. pp. 192-194

[27] Webster AD, Quinlan JD, Richardson PJ. The influence of paclobutrazol on growth and cropping of sweet cherry cultivar. The effect of annual soil treatments on the growth and cropping of cultivar 'early rivers'. Journal of Horticultural Science. 1986;**61**(4):471-478

[28] Gaash D. Growth retardation of apple, plum and apricot trees by paclobutrazol in a Mediterranean climate. Acta Horticulturae. 1986;**2**:559-562

[29] Looney NE, Mckeller JE. Effects of foliar and soil surface applied paclobutrazol on vegetative growth and fruiting quality of sweet cherries. Journal of the American Society for Horticultural Science. 1987;**112**(1):71-76

[30] Blanco A. Control of shoot growth of peach and nectarine trees with paclobutrazol. Journal of Horticultural Science. 1988;**63**(2):201-207

[31] Kaska N, Kuden A, Kuden AB. Physiological effects of PP333 (paclobutrazol) on golden delicious apple cultivars. Journal of C.U. Agricultural Faculty. 1991;**5**(4):87-94

[32] George AP, Nissen RJ, Campbell JA, Rasmussen T, Allan P, Shaltow AD. Effects of paclobutrazol on growth and

yield of low chill peaches in subtropical Australia. Acta Horticulturae. 1995;**409**:109-116

[33] Kuden A, Kuden AB, Naska N. Physiological effects of foliage applied paclobutrazol on Canino and Precoce de Colomer apricot cultivars. Acta Horticulturae. 1995;**384**:419-423

[34] Sharma MK, Joolka NK. Influence of tricontanol and paclobutrazol on the drought tolerance of non-Pareil almond. Indian Journal of Horticulture. 2000;**57**(2):118-122

[35] Arzani K, Roosta HR. Effects of paclobutrazol on vegetative and reproductive growth and leaf mineral content of mature apricot (*Prunus armeniaca* L.) trees. Journal of Agricultural Science and Technology. 2004;**6**:43-55

[36] Pant N, Ratan K. Effect of paclobutrazol and chloromequat on growth, flowering, yield and quality of 'Red Delicious apple. Progressive Horticulture. 2004;**36**(1):167-170

[37] Wani WM. Influence of growth retardant (paclobutrazol) and gibberellic acid on growth and production of sweet cherry [PhD thesis]. Srinagar, India: Sher-e-Kashmir University of Agricultural Sciences and Technology of Kashmir; 2004. Submitted

[38] Gupta NK, Bist LD. Effect of different planting systems and paclobutrazol on vegetative growth of Bagugosa pear. Indian Journal of Horticulture. 2005;**62**:20-23

[39] Asin L, Alegre S, Montserrat R. Effect of paclobutrazol, prohexidene-Ca, deficit irrigation, summer pruning and root pruning on shoot growth, yield and return bloom in a 'Blanquilla' pear orchard. Scientia Horticulturae. 2007;**113**:142-148

[40] Sharma MK, Joolka NK. Influence of triacontanol and paclobutrazol on growth and leaf nutrient status of non Pareil almond under different soil moisture regimes. Indian Journal of Horticulture. 2011;**68**(2):180-183

[41] Abdollahi M, Eshghi S, Tafazzoli E, Moosavi N. Effects of paclobutrazol, boric acid and zinc sulfate on vegetative and reproductive growth of strawberry cv. Selva. Journal of Agricultural Science and Technology. 2012;**14**:357-363

[42] Mir MM, Baba JA, Umar I, Rather GH, Rehman MU, Banday SA. Effect of soil applied paclobutrazol on vegetative growth and quality attributes of apricot (*Prunus armeniaca* L.). Green Farming International Journal. 2015;**6**(4):813-813

[43] Ashraf N, Bhat MY, Sharma MK, Ashraf M. Influence of paclobutrazol and summer pruning on growth and vigour of apple cv. Red delicious. Journal of Food, Agriculture and Environment. 2015;**13**(2):98-100

[44] Ashraf N, Bhat MY, Sharma MK, Rather GH, Ashraf M, Dar MA, et al. Effect of paclobutrazol and summer pruning on yield and fruit quality of apple cv. 'Red Delicious'. Applied Biological Research. 2015;**17**(2):166-173

[45] Hedden P, Graebe JE. Inhibition of gibberellins biosynthesis by paclobutrazol in cell free homogenates of *Cucurbita maxima* endosperm and *Malus pomila* embryos. Journal of Plant Growth Regulators. 1985;**4**:111-112

[46] Webster AD, Quinlan JD. Chemical control of tree growth of plum (*Prunus domestica* L.). preliminary studies with the growth retardant paclobutrazol (PP_{333}). Journal of Horticultural Science. 1984;**59**(3):367-375

[47] Martin GC, Yoshikawa F, James HL. Effect of soil application of

paclobutrazol on vegetative growth, pruning time, flowering, yield and quality of 'Flavorcrest' peach. Journal of the American Society for Horticultural Science. 1987;**112**(6):915-921

[48] Stan S, Burloi N, Popescu I, Feneanu N, Cotorobai M. Performance of paclobutrazol (Cultar) in controlling vegetative growth and cropping of stone fruits. Acta Horticulturae. 1989;(239):221-228

[49] Wolstenholme N, Whiley W, Saranah JB. Manipulating vegetative and reproductive growth in avocado *(Persea americana* mill.) with paclobutrazol foliar sprays. Scientia Horticulturae. 1990;**41**:315-327

[50] Jindal KK, Chandel JS. Effect of triacontanol and paclobutrazol on fruit set, growth and quality of *Prunus saliciana*. Indian Journal of Horticulture. 1996;**53**(4):262-268

[51] Arzani K, Wood DES, Lawes GS. Influence of first season application of paclobutrazol, root-pruning and regulated deficit irrigation on second season flowering and fruiting of mature Sundrop apricot trees. Acta Horticulturae. 2000;**516**:75-82

[52] Davenport TL. Reproductive physiology of mango. Brazilian Journal of Plant Physiology. 2007;**19**(4):363-376

[53] Yadav RK, Rai N, Yadav DS, Asati BS. Use of paclobutrazol in horticultural crops—A review. Agricultural Reviews. 2005;**26**(2):124-132

[54] Abdel Rahim AOS, Elamin OM, Bangerth FK. Effects of paclobutrazol (pbz) on floral induction and associated hormonal and metabolic changes of biennially bearing mango (*Mangifera indica* L.) cultivars during off year. Journal of Agricultural and Biological Science. 2011;**6**(2):55-67

[55] Abdel Rahim AOS, Elamin OM, Bangerth FK. Effects of paclobutrazol on floral induction and correlated phyto-hormonal changes in grafted seedlings of different mango (*Mangifera indica* L.) cultivars. Sudan Journal of Agricultural Research. 2008;**11**:111-120

[56] Blaikie SJ, Kulkarni VJ, Muller WJ. Effects of morphactin and paclobutrazol flowering treatments on shoot and root phenology in mango cv. Kensington pride. Scientia Horticulturae. 2004;**101**:51-68

[57] Jose ARS, Reboucas TNH. Use of paclobutrazol in mango orchard in southwest region, Bahia state, Brazil. Acta Horticulturae. 2000;**509**:713-715

[58] Carreno J, Oncina R, Carreno I, Tornel M. Effect of paclobutrazol on vegetative growth, grape quality and yield of Napoleon table grape variety. Acta Horticulturae. 2007;**754**:179-182

[59] Gaurang PD, Patel BN, Desai KD, Patel NK, Patel BB. Influence of paclobutrazol for earliness in mango cv. Alphonso. International Journal of Science, Environment and Technology. 2016;**5**(5):2713-2718

[60] Huang WD, Shen T, Han ZH, Liu S. Influence of paclobutrazol on photosynthesis rate and dry matter partitioning in the apple tree. Journal of Plant Nutrition. 1995;**18**:901-910

[61] Kumar A, Singh CP, Ram S. Paclobutrazol influences post harvest life of mango Cvs. Dashehari, Langra, Chausa and Fazri. Journal of Plant and Environmental Research. 2019;**3**:13 (1-11)

[62] George AP, Nissen RJ. Effects of water stress, nitrogen and paclobutrazol on flowering, yield and fruit quality of the low-chill peach cultivar, 'Flordaprince'. Scientia Horticulturae. 1992;**49**(3):197-209

[63] Asin L, Vilardell P, Carrera M. Opciones de la combinacion de poda de raices con Paclobutrazol o Prohexadiona de Calcio en la reduccion del vigor de plantaciones de pera Blaquilla y Conference. Actas Portuguesas de Horticultura. 2005;**6**(2):36-41

[64] Wani AM, Peer FA, Lone IA. Effect of paclobutrazol on growth, picking maturity and storage behaviour of Red Delicious apples. The Asian Journal of Horticulture. 2007;**2**(1):171-175

[65] Bill M. Advancing harvest maturity and improving storability of 'Triumph' Persimmons [M.Sc. thesis]. Stellenbosch University; 2012. pp. 119

[66] Loreti F, Scalabrelli G, Musso O. Effect of paclobutrazol on productive behaviour of peach under protected cultivation. Acta Horticulturae. 1989;**239**:285-288

[67] Costa G, Sabatini E, Spinelli F, Andreotti C, Bomben C, Vizzotto G. Two years of application of Prohexadione-Ca on apple. Effect on vegetative and cropping performance, fruit quality, return bloom and residual effect. Acta Horticulturae. 2004;**653**:35-40

[68] Khader SESA. Orchard application of paclobutrazol on ripening, quality and storage of mango fruits. Scientia Horticulturae. 1990;**41**(4):329-335

[69] Ben-Arie R, Zhou HW, Sonego L, Zutkhi Y. Plant growth regulator effects on the storage and shelf life of 'Triumph' persimmon. Acta Horticulturae. 1997;**436**:243-250

[70] Erez A. Dwarfing peaches by pruning and by paclobutrazol. Acta Horticulturae. 1984;**146**:235-241

[71] Blanco A. Effects of paclobutrazol and ethephon on cropping and vegetative growth of 'crimson gold' nectarine trees. Scientia Horticulturae. 1990;**42**:68-73

[72] Greene DW. Reduced rate and multiple sprays of paclobutrazol control growth and improve fruit quality of delicious apples. Journal of the American Society for Horticultural Science. 1991;**116**:807-812

[73] Shakeri F, Baninasab B, Ghobadi C, Mobli M. Effect of paclobutrazol concentration and application methods on vegetative and reproductive growth of strawberry (*Fragaria x ananassa* Duch. cv. Selva). Journal of Horticultural Science. 2009;**23**:18-24

[74] Elfving DC, Lougheed EC, Chu CL, Cline RA. Effects of diaminozide, paclobutrazol and uniconazole treatments on 'McIntosh' apples at harvest and following storage. Journal of the American Society for Horticultural Science. 1990;**115**:750-756

[75] Sarker BC, Rahim MA, Archbold DD. Combined effects of fertilizer, irrigation and paclobutrazol on yield and fruit quality of mango. Horticulturae. 2016;**2**(4):14

[76] Burondkar MM, Jadhav BB, Chetti MB. Effect of plant growth regulators, polyamine and nutrients on fruit quality and occurrence of spongy tissue in alphonso mango. Proceedings of the VIIIth International Mango Symposium. Acta Horticulturae. 2009;**820**:689-696

[77] Rebolledo-Martinez A, del Angel-Perez AL, Moreno JR. Effects of paclobutrazol and KNO_3 over flowering and fruit quality in two cultivars of mango manila. Interciencia. 2008;**33**(7):518-522

[78] Ashraf N, Ashraf M, Bhat MY, Sharma MK. Paclobutrazol and summer pruning influences fruit quality of Red

Delicious apple. International Journal of Agriculture, Environment and Biotechnology. 2017;**10**(3):349-356

[79] Nilsson T, Gustavsson KH. Postharvest physiology of aroma apples in relation to position on the tree. Postharvest Biology and Technology. 2007;**43**:36-46

[80] Elfving DC, Chu CL, Lougheed EC, Cline RA. Effects of diaminozide and paclobutrazol treatments on fruit ripening and storage behaviour of 'McIntosh' apple. Journal of the American Society for Horticultural Science. 1987;**112**:910-915

[81] Luo Y, Yainwright H, Morre KG. Effects of orchard application of Paclobutrazol on the composition and firmness of apple fruits. Scientia Hori. 1987;**39**:301-309

[82] Huder DM. The use of fertilizers and organic amendments in the control of plant disease. In: Pimentel D, editor. Handbook of Pest Management in Agriculture. Vol. 1. Boca Raton, FL: CRC Press; 1981. pp. 357-394

[83] Bangerth F, Dilley DR, Dewey DH. Effect of postharvest calcium treatments on internal breakdown and respiration of apple fruits. Journal of the American Society for Horticultural Science. 1972;**97**:679-682

[84] Rieger M. Paclobutrazol, root growth, hydraulic conductivity and nutrient uptake of 'Nemaguard' peach. Horticultural Science. 1990;**25**(1):95-98

[85] Allan P, George AP, Nissen RJ, Rasmussen TS, Shaltout AD. Phenology and management practices of 'Flordaprince' peach. Acta Horticulturae. 1995;**409**:19-24

[86] Huett DO, George AP, Slack JM, Morris SC. Diagnostic leaf nutrient standards for low-chill peaches in subtropical Australia. Australian Journal of Experimental Agriculture. 1997;**37**(1):119-126

[87] Bhatia SK. Effect of paclobutrazol, daminozide and rootstocks on the growth and chemical composition of apple plant [PhD thesis]. Nauni, Solan, India: Dr. Y.S. Parmar University of Horticulture and Forestry; 1992

[88] Curry EA. Chemical control of vegetative growth of deciduous fruit trees with paclobutrazol and RSWO 411. Horticultural Science. 1988;**23**:470-473

[89] Lichou J, Devlavnay V, Garcin A, Vidaud J, Jay M. Application of paclobutrazol to stone fruit trees. In: Effects on Vegetative Growth and Fruiting. France: CTIFL-INRA; 1988. pp. 215-217

[90] Swietlik D, Miller SS. The effect of paclobutrazol on the growth and uptake of Ca and K by apple seedlings. Journal of Plant Nutrition. 1984;**7**:1555-1556

[91] Bonomo R, Neri D, Sansavini S. Dry matter and nutrient partioning, seasonal water consumption and gas exchange of paclobutrazol growth-controlled apple trees. Acta Horticulturae. 1989:239, 159-166

[92] Swietlik D, Miller SS. The effect of paclobutrazol on mineral nutrition of apple seedlings. Journal of Plant Nutrition. 1985;**8**:369-382

Chapter 4

Flowering of Sweet Cherries "*Prunus avium*" in Tunisia

Thouraya Azizi-Gannouni and Youssef Ammari

Abstract

In Tunisia, the development of cherry growing is limited by two major constraints, namely, the chilling requirements and the self-incompatibility of some cultivars. In order to contribute to the development of this high added-value culture, which is capable to play an important socioeconomic role in rural and semi-forestry places, this study has set the main objective, characterization, and selection of best-suited cultivars to mild winter based on the blooming period. The plant materials used for this study are composed of the introduced cultivars, which are "Napoleon," "Van," "Moreau," "Sunburst," and "Stella," and unknown cultivars, which are "V1," "V2," "V3," "V4," and "V5," and a local one "Bouargoub." Differential behavior between cultivars was shown for phenological stages (budbreak, flowering, maturity, and leaf fall), and this behavior is dependent in some cases on the cold requirement [chilling requirements (CR)]. The local cultivar "Bouargoub" recorded the lowest "CR" with early flowering and maturity.

Keywords: behavior, chilling requirement, flowering, mild winter

1. Introduction

Sweet cherry tree is a hardy species capable of adapting to various soil and climatic conditions; but its development is surrounded by climatic and physiological constraints. The sweet cherry (Prunus avium L.) from Rosaceae family, with a number of chromosomes (2n = 2x = 16), is an allogamous species that is adopted to a self-incompatibility system to ensure cross-fertilization. Self-compatibility in sweet cherry occurs rarely in nature and, consequently, there are a reduced number of self-compatible cultivars [1]. Sweet cherry is the first fruit of summer season and is highly appreciated by consumers, cultivated for its edible fruits and its wood.

According to FAO [2], world production of sweet cherries has been estimated at 2,245,826 tones. The largest cherry-producing countries are Turkey, the United States, Iran, Italy, and Spain [2]. The cultivation of sweet cherries in Tunisia covers an area of about 961 ha [3]. This species is particularly cultivated in the region of northern Tunisia, where the winter is mild and spring frosts are rare. National production is estimated at 5187 tons [3]. Despite the favorable conditions, sweet cherry is poorly valued and only some regions practice this culture in small-scale along Tunisia.

The local cultivar is poorly valued, given the predominance of introduced cultivars which have high productivity and good adaptation to the mild North African climate. The characterization of this variety and the comparison with other introduced varieties is essential for its conservation. In this context, the present work is focused on the study of the phonological characteristics such as blooming stage of some cultivars of sweet cherry trees in relation to the conditions of the environment. Flowering is a determining factor in fruit production based on pollen self-incompatibility. The cultivation of cherry trees is limited in the north of Tunisia in the high altitudes to fill chill requirement. Therefore, this study is part of the evaluation, the development of genetic resources in fruit arboriculture in Tunisia, and the extension of the cultivation of cherry trees in regions at medium altitude.

For sweet cherry, like for other temperate-zone fruit species, when chilling requirements are not adequately satisfied, negative repercussions on productivity occur. Insufficient chilling can lead to erratic, delayed budbreak, and heterogeneous flowering. Chilling increased the flower size, pedicel length, and fruit set [4].

In many perennial species, it has been shown that increase in temperatures during the last dormant stage (autumn, winter) was responsible for advancing blooming dates, leading to an increased risk of damage caused by late frosts, phenological disorders, with a large spread of flowering dates and difficult synchronization of flowering with the activity of pollinators.

In a Mediterranean climate and specifically in Tunisia, development of sweet cherry growing (Prunus avium) shows several problems related to floral biology, chilling requirement, appearance of bud anomalies, and inconstant and extremely low yield.

The aim of our study was to investigate the blooming phenophase and the effect of temperature during flowering period on the fruit set and the production of 11 cultivars of sweet cherry in the climate condition of Tunisia, from which researchers and orchard managers will get reliable information for their study or planting.

2. Experimental sites and plant materials

The behavior of the different cultivars was monitored in three experimental sites located in three regions of northwestern Tunisia with different pedo-climatic characteristics:

The site of Ain-Draham is located at latitude 36°46′34″ North and longitude 8°41′05″ East, with an altitude of 800 m. The average annual rainfall was about 1040 mm. The lowest average temperature was about 6.08°C during February and the warmest was 26°C during July. The bioclimatic floor is humid superior with temperate winter. The site of Bousalem is at latitude 36°36′34″ North and longitude 8°58′17″ East, with an altitude of 127 m above sea level. The average annual rainfall was about 57.24 mm. The lowest average temperature was around 10.03°C during February and the highest temperature was 35.50°C during July. The bioclimatic floor is subhumid with temperate winter. The site of Tibar is at latitude 36°31′21″ North and longitude 9°06′22″ East, with an altitude of 328 m. The average annual rainfall was about 540 mm. The lowest average temperature was about 8.37°C during February and the hottest temperature was around 29°C during July–August. The bioclimatic floor is subhumid with mild winter.

The plant materials used in this study are composed of 11 cultivars of local and introduced sweet cherries (Prunus avium L.) of known and unknown origins. These cultivars are unequally distributed between the three experimental sites (**Table 1**).

The different studied flower traits are the length of the pistil (LPIST), the ovary area (SROV) and the number of stamens (NBET), length (Lopt) and width of petals (Larpt) and flower diameter (DFL), shape of petals (SHPE), and the arrangement of petals (ARPE).

		Sites				
Cultivar	**Origin**	**Ain-Draham**	**Bousalem**	**Tibar**	**(In) compatibility groups***	**S allele composition***
Napoléon	Germany	+	—	+	III	S3S4
Van	Canada	+	—	+	II	S1S3
Moreau	French	+	—	+	XVI	S3S9
Sunburst	Canada	+	—	+	**SC	S3S4'
Stella	Canada	+	—	—	**SC	S3s4'
Bouargoub	Tunisia	+	—	—	*XLII	S2S10
V1 unknown	—	—	+	—	XVI	S3S9
V2 unknown	—	—	+	—	**SC	S3S4'
V3 unknown	—	—	+	—	XVIII	S1S9
V4 unknown	—	—	+	—	**SC	S3S4'
V5 unknown	—	—	+	—	II	S1S3

+, indicates that the variety is tested in the relevant site.
**The S-genotype and incompatibility groups [5] according to Schuster [6].*
***SC: Self Compatible.*

Table 1.
Name, origin, distribution, and S-genotype of the studied cultivars per experimental site.

The flowers were collected in full bloom by using five flowers per tree on five trees by cultivar and site. The different measurements were carried out using a vernier caliper for measuring the length and width of petals and flower diameter. However, the ovary area, pistil length, and number of stamens were carried out with electronic scanning microscope (Leica).

- Statistical analyses were performed using SAS 9.1. ANOVA was carried out and means were separated by the LSD test ($\alpha \leq 0.05$).

3. Flowering of sweet cherry

The transition from the vegetative state to a reproductive state is a crucial stage of development in fruit trees, and this transition is marked by floral induction. During the vegetative phase, the vegetative meristems produce leaves and stems necessary for the accumulation of sufficient reserves to eventually lead to the growth of the tree depending on its genotype and environmental conditions [7].

These meristems become inflorescential, producing flowers. The success of this sexual reproduction depends both on the sufficient accumulation of reserves and on a synchronous reproductive phase with optimal environmental conditions for flowering and fruiting [8].

The response of flowering at room temperature is variable depending on the species and genotypes. Studies carried out on different accessions of *Arabidopsis* have shown that high temperature favors flowering [9], which leads to the conclusion that flowering is dependent on warm temperatures.

The trunk and branches carry spurs (**Figure 1**) called "bouquets of May" because their development is generally completed at the end of May. The flowers appear in all cases at the base of the annual shoots of the previous year, whether it is long shoots of the trunk and branches or bouquets of May. On a cherry tree a

Figure 1.
Spurs (bouquet of May) in Ain-Draham site.

few years old, most of the flowering intended for fruit production is carried by the bouquets of May. The good development of these bouquets is very important to maintain a good quality of cherries production [10].

The flower of this genus is generally characterized by the following features: flower with five petals and five sepals, solitary carpel with a terminal style [11]. It is a hermaphrodite flower and the fruit is a drupe [12]. These drupes are most often edible and delicious but sometimes bitter or sour (cherries, sloes), more rarely toxic (fruits of the cherry laurel).

The development of flower buds is under biochemical control. This biochemical signal allows the tissue to change from the vegetative state to the reproductive state [13]. It occurs due to a balance between gibberellic acid, auxin, cytokinins, and ethylene-type hormones [14]. The floral initiation (sepals, petals, stamens, and pistil) of sweet cherry occurs after harvest [14].

3.1 Different characteristics of flowers traits in relation to the environment conditions

The morphology of fruit species provides information on the adaptation and behavior of these species with regard to their environmental conditions. Indeed, the size of the flowers is generally considered to be the most important factor for pollinators.

For each experimental site, the results of the multiple comparisons of means for the different flower traits (**Figure 2**) are presented in **Table 2** and in the study of Azizi-Gannouni et al. [15].

Azizi-Gannouni et al. [15] showed significant variability in the number of stamens for the same variety between the two different sites (Ain-Draham and Tibar). The local cultivar "Bouargoub" has a longer pistil compared to other cultivars, while V4 (Bousalem site) recorded the shortest pistil (**Table 2**). Genotypic differences that control the dependence of these floral parameters on its genetic potential are to be excluded in view of the different behaviors of the same cultivar in pedoclimatically different experimental sites [15] (**Figure 3**).

Flowers with large diameters generally attract more pollinators [16]. This directly affects the pollination of flowers and therefore their setting and their production. Morphological monitoring carried out on all the cultivars in the three sites shows that in Bousalem site the flowers of small diameter have a high fruit production (**Figure 4**), which contradicts the results of the work of Johnson et al. [17] and Wetzstein et al. [18].

Figure 2.
The different studied flower traits of cultivars in Ain-Draham site.

Sites	Cultivars	DFL (mm)	Lopt (mm)	Larpt (mm)	SHPE	ARPE
Ain-Draham	Napoléon	34.60b	13.8b	12.2c	Circular	Intermediate
	Van	28.40c	11.2c	11.6c	Circular	Intermediate
	Moreau	25.80c	11.4c	10.2d	Broad obovate	Disjoints free
	Sunburst	40.40a	18.2a	11.8c	Circular	Disjoints free
	Stella	41.60a	18.8a	14.94b	Medium obovate	Disjoints free
	Bouargoub	41.66a	19.33a	17.96a	Broad obovate	Overlapping
Tibar	Napoleon	44.12a	19.56a	18.9a	Circular	Intermediate
	Van	38.2b	16.6b	15.2c	Circular	Intermediate
	Moreau	39b	17.5b	16.78b	Broad obovate	Disjoints free
	Sunburst	29.16c	12.58c	9.78d	Circular	Disjoints free
Bousalem	V1	41a	18a	17a	Circular	Intermediate
	V2	33.6b	14.8b	14.7b	Medium obovate	Disjoints free
	V3	43.96a	19.48a	15.28b	Circular	Intermediate
	V4	32.04b	15.02b	12.9c	Medium obovate	Disjoints free
	V5	43.8a	18.9a	17.26a	Broad obovate	Overlapping

Different small letters in the same column indicate significantly different values within cultivars at $\alpha \leq 0.05$.

Table 2.
*Mean of the parameters measured on sweet cherry flowers (*Prunus avium *L.) at three sites.*

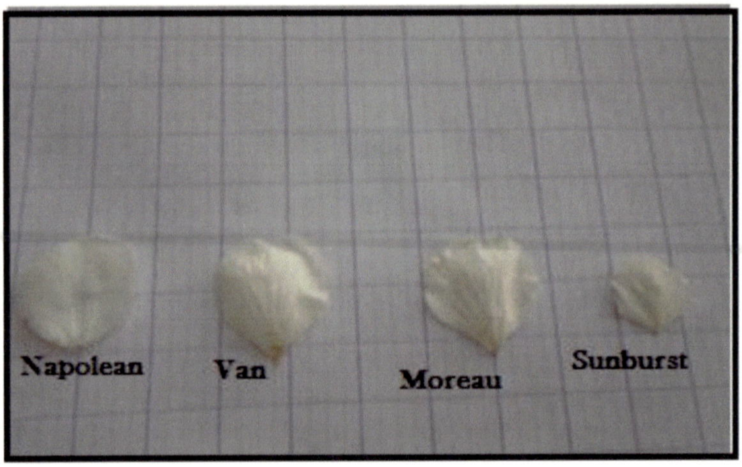

Figure 3.
*Shape of Petals of the four cultivars (*Prunus avium *L.) in Tibar site.*

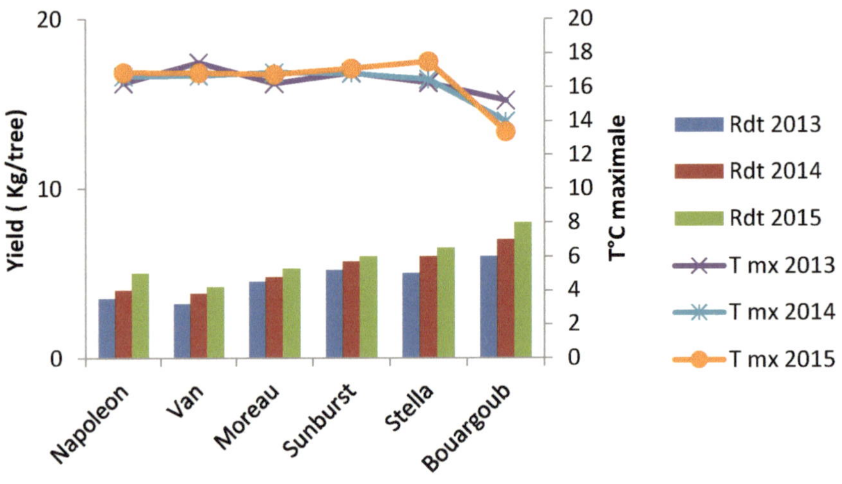

Figure 4.
Yield (Kg/tree; Rdt) and Maximum temperature during blooming at Ain-Draham site.

Furthermore, in the three study sites, the collected data show a significant effect of the site on the morphology and size of all the tested cultivars (identified and unidentified). This confirms the important effect of climatic conditions on this parameter, a result confirmed by the work of Niu et al. [19], who showed that the diameter of the flower is more influenced by daytime temperatures than by night-time temperatures and this dimension has no relation to the difference of temperatures between day and night.

Whereas, in the Ain-Draham site, it is observed that the local cultivar "Bouargoub" has the largest flower diameter compared to "Van" and "Moreau" cultivars (**Table 2**). Likewise, it has the longest style but the smallest ovary surface with an intermediate number of stamens [15].

Lu [20] has shown that cultivars grown in a warm winter climate give flowers with longer styles than these grown in a cold winter climate, which contradicts the results of the present work (**Table 2**), where the cultivar "Napoleon" shows a longer style in Ain-Draham (climate with cold winters) than in Tibar (climate with mild winters).

	DFL	Lopt	Larpt
Site	303.92***	363.73***	854***
Cultivars	102.74***	63.61***	235***
Site*cultivars	334.71***	348.13***	258***

NS, not significant.
**Significant at 0.05.*
****Significant at 0.001.*

Table 3.
Inter-site variance analysis: Values and significance of the F-test for flower traits.

Referring to the hypothesis relating the floral diameter to pollen self-compatibility, our results show the invalidity of this assumption since the self-compatible cultivars "Sunburst," "Stella," and "Bouargoub" show large floral diameters as well as self-incompatible "V1," "V3," and "V5" [5].

The shape and arrangement of the petals do not change according to the pedo-climatic conditions of the experimental site, and it can be said that these two morphological traits depend on the genetic potential of the cultivar.

3.2 The effect of the environment and genotype on the floral parameters

The variance analysis (**Table 3**) makes it possible to test for the different flower parameters, the effect of the cultivars, the effect of the sites, and the effect of the interaction "cultivar x sites." The inter-sites' analyses were done for the four com-moncultivars (Napoleon, Van, Moreau, and Sunburst) at the two experimental sites of Tibar and Ain-Draham.

The effect of the interaction "cultivars x sites" was significant ($p < 0.05$ to $p < 0.001$) for the morphometric measured characteristics for the flower (**Table 3**).

The "Genotype × Environment" interactions were observed for flowers' quantitative parameters [15] (**Table 2**). This indicates that the sources of variation are both genetic and environmental. In addition, the differences in altitude and soil-climatic conditions of the two study sites may explain the observed variations. In our study, all of the cultivars cultivated at the three sites showed the inter-varietal variability in flower traits and therefore these three geographical areas should be the priority sites for in situ conservation.

4. Climatic change and chilling requirement

Global warming, the phenomenon of sustainable rise in ocean and atmospheric temperatures, is the main form of climate change. Terrestrial temperature measurements made during the twentieth century show an increase in the average temperature. This warming would have taken place during the twentieth century in two phases, the first from 1910 to 1945, and the second from 1976 to the present [21].

Human activities are therefore the dominant cause of the warming observed over the last 50 years on Earth [22]. This climate change is already having consequences on the biodiversity and ecosystems [8]. Temperature is an influencing factor on the development and growth of plants. Climate change can therefore have a major impact on their phenology. Changes in the phenological stages such as the date of leaf coloring [23] and blooming [24] have already been observed. The advance of the growing period has been linked to climate [25, 26].

Phenology is the main biological parameter of climate change and is one of the main key characteristics of the adaptability of species to these changes.

The exposure to these cold temperatures and the satisfaction of chilling requirements is necessary in several species by the resumption of growth in the spring. Predicting the break of dormancy in fruit trees is essential for producers. Knowing the date of budding makes it possible to estimate the length of the growing season and the risk of frost damage. Global warming can cause a decrease in the number of chill units for certain regions, which will have an impact on the date of bud burst [27]. A limited supply of chill units decreases fruit production [28].

4.1 Chill accumulation in the three studied sites

Sweet cherry trees develop their vegetative and fruiting buds in summer. As winter approaches, the already developed buds remain dormant to protect themselves from the cold. These buds remain dormant until they have accumulated sufficient chill units. They break up in response to high temperatures and following a sufficient accumulation of chill. If the buds do not receive a sufficient chilling requirement during the winter, the trees will develop one or more of the physiological symptoms such as heterogeneous and spreading flowering, a reduction in the quality of fruit (degree of firmness, size of the fruit) and the fruit set rate.

Table 4 shows the mean chilling accumulation registered in Ain-Draham, Bousalem, and Tibar from October 1 to March 1 during the three consecutive years (2012–2013, 2013–2014, and 2014–2015). The chill accumulation is expressed in chill units (CU) (Utah model), chill portions (CP) (Dynamic model), and hours below 7°C (Weinberger model). A noteworthy difference between chill accumulations in three experimental areas was found using any of the three described models.

Under field conditions, the coefficients of variation between October 1 and March 1 during the 3 years at Bousalem were relatively high when using the Utah and Dynamic models (CV = 13.98, 5. 42%, respectively), which indicates that the chill accumulation varies from year to year (**Table 4**). The Ain-Draham station presented CV values slightly lower than those of Tibar. Using the three models, chill accumulation is low in Bousalem, intermediate in Tibar, and significantly higher in Ain-Draham.

The three studied areas registered a different chill accumulation explained by altitude location which is in accordance with results obtained by Alburquerque et al. [29] and geographic distance between sites. Bousalem is at a lower altitude (127 m above sea level). Nevertheless, Ain-Draham is at a higher altitude (800 m above sea level). Tibar is at an intermediate altitude (328 m above sea level).

	Dynamic model			Utah model			Weinberger model		
	Mean (CP)	CV%	SD	Mean (CU)	CV%	SD	Mean (H < 7°C)	CV%	SD
Ain-Draham	80.17	2.32	1.86	1840	3.79	69.77	1044	4.36	45.61
Tibar	62.82	2.99	1.88	1014	4.21	42.71	480*	10.42	50.08
Bousalem	55.11	5.42	2.9	767.33*	13.98	107	296.33	3.6	10.96

**Significant at ≤0.05.*

Table 4.
Chill accumulation in the period November-March between 2012 and 2015 in Ain-Draham, Bousalem, and Tibar. Results are expressed in chill units (Utah model), chill portions (dynamic model), and hours below 7°C(Weinberger model).

The Bousalem site is characterized by a mild winter with less chill accumulation calculated according to the three models. The Tibar site is milder than Ain-Draham with less average chill accumulation (**Table 4**).

4.2 Chilling requirement for breaking dormancy

From the beginning of the chilling accumulation (first week of November), five branches of each cultivar (length of 40 cm, base diameter of 8–10 mm) were picked every 3–4 days from trees in the orchards and the bases were placed in a 5% sucrose solution in a growth chamber, making a fresh cut at the base of the branches [30, 31].

The branches were maintained at 25 ± 1°C under white fluorescent tubes (55 mol $m^{-2} s^{-1}$) with a photoperiod of 16 h and at 18 ± 1°C during a dark period of 8 h, with a constant relative humidity of 70%. The sucrose solution was refreshed and changed every 5 days. Branches were maintained for 10 days for forcing in the growth chamber. The date of breaking dormancy was established when, after 10 days in the growth chamber, 30% of the flower buds had reached the phenological growth stage 53–55 (**Figure 5**) according to the international BBCH scale [32]. The chilling requirements (CRs) coincided with the chilling accumulated until the date of dormancy release.

The chilling requirements for breaking dormancy of the sweet cherry cultivars planted in Ain-Draham, Tibar [31], and Bousalem (**Table 5**) showed different chilling requirements (CR) compared to the geographic area and climatic conditions of the year according to the three models. The Dynamic model is used to determine the chill requirements of different cultivars since it is the adequate model for Mediterranean conditions [29].

Figure 5.
Stage 53 (Bud burst) [30].

		V1	V2	V3	V4	V5
Bousalem	Mean	23.33b	50.66a	26b	53.66a	54.33a
	CV	24.74	6.02	19.98	5.69	6.46

Different small letters in the same row indicate significantly different values within cultivars at α ≤ 0.05.

Table 5.
Chilling requirements (mean; coefficient of variation, %) for breaking of dormancy expressed in chill portions (CP) for the cultivars in Bousalem site.

These results and the study of Azizi-Gannouni et al. [31] showed that the cultivars "Bouargoub," V1, and V3 registered less chill requirements than the other cultivars.

If we compare our results with those found by Alburqurque et al. [29] in Murcia (southeastern Spain), we can find some cultivars close to that cultivated in Bousalem using the Dynamic model. "V2" required the same chill requirements (48 CP) as "Ruby," "Somerst," and "Burlat." "V4" required the same CP as New Star (53.5 CP). "V1" and "V3" were almost close to Cristobalina (30 CP).

According to the three models, the cultivars "V1" and "V3" do not need a large amount of chill and are better favored in the north of Tunisia. However, "V2," "V4," and "V5" can be grown in this region provided they meet their chilling requirements (CR). Our results suggested that chilling requirements are the main factor for determining the date of flowering in sweet cherry. The date of flowering of the cherry tree in the north of Tunisia was influenced by the cold rather than by the heat and probably, by other biochemical factors of the plant.

In terms of low chilling requirement, "V1" and "V3" were the best cultivars, but they recorded the lowest yield. However, "V2" and "V4" need to accumulate a large amount of chill (CR) to register the highest fruit yield. "V5" was poorly adapted to the North Tunisian climate. It needs large chilling requirements and it has generated the lowest yield. For future improvement programs, we can choose "V1" and "V3" for their low chilling requirements, "V4" and "V2" for their high yields.

5. Determination of flowering date

Phenological monitoring of flowering was carried out on five trees per cultivar per site during the 2012–2013, 2013–2014, and 2014–2015 seasons. Flowering (**Figure 6**) was observed from mid-March to mid-April depending on the region, cultivars, and climatic conditions. The start of blooming was taken as the day on which 10% of the flowers on the tree were opened, full blooming was when 75% of the flowers were opened, and the end of blooming was when 95% of the petals fell [33]. Periodical checks (every 2–3 days) were carried out on the trees for this purpose.

The graphical representations of the different phases of flowering for the three sites during the years 2012–2013, 2013–2014, and 2014–2015 are shown in **Figures 7–9**, respectively.

Figure 6.
Blooming of "Bouargoub" cultivar.

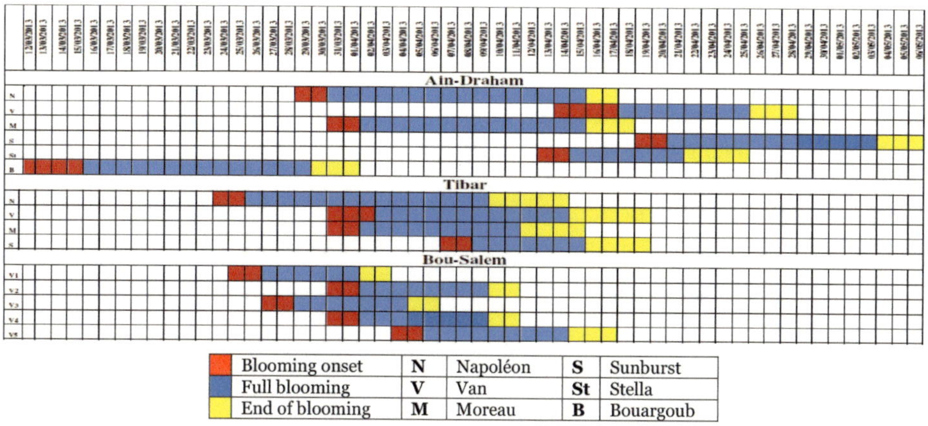

Figure 7.
Spreading of sweet cherry blooming at the three sites during 2013.

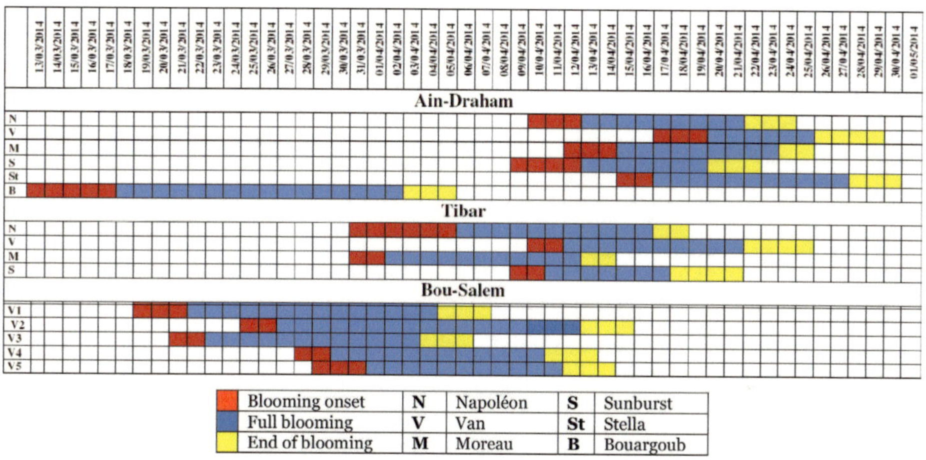

Figure 8.
Spreading of sweet cherry blooming at the three sites during 2014.

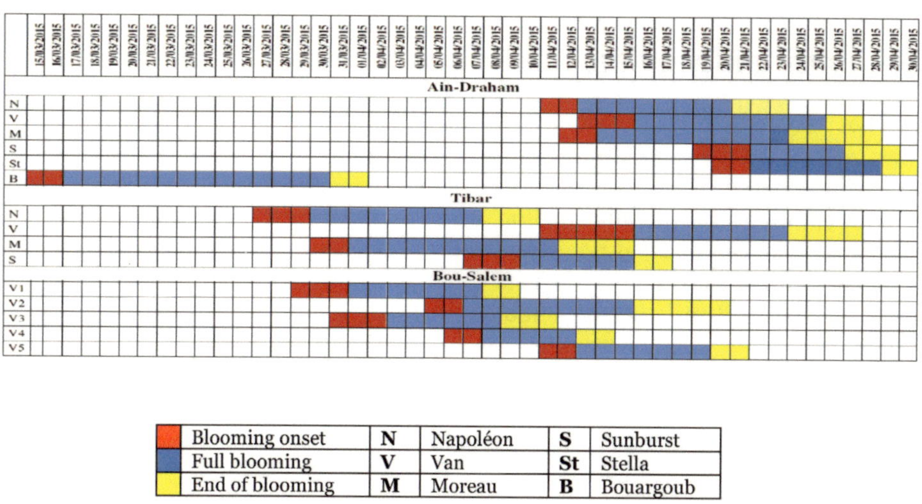

Figure 9.
Spreading of sweet cherry blooming at the three sites during 2015.

5.1 Blooming in Ain-Draham site

The date of blooming of the different cultivars in the Ain-Draham site is offset from Tibar and Bousalem. The local cultivar "Bouargoub" showed an early flowering followed by "Napoleon" during the 3 years of study. The blooming period for "Bouargoub" is more spread out than the other cultivars (18–24 days).

In 2015, the blooming period was reduced, and it was between 11 and 17 days for "Stella" and "Moreau," respectively. The blooming period of all cultivars was reduced during 2015 with the exception of "Bouargoub" which keeps the same period as 2013. "Van" has the same blooming period during 2013 and 2015, with a shortening of 2 days during 2014 (**Figures 7–9**).

5.2 Blooming in Tibar site

Blooming was advanced in the Tibar site compared to the Ain-Draham site for the same cultivars and during the 3 studied years. "Sunburst" was characterized by the shortest blooming period and a moderately late start to blooming, while "Napoleon" had the longest period between 15 and 22 days and an early blooming start.

With a high monthly temperature in Tibar, blooming started earlier than in Ain-Draham. Observations of the blooming periods, "Napoleon," "Van," and "Moreau," registered longer periods than that in Ain-Draham. "Sunburst" kept almost the same blooming period (13 days) in 2013 and 2014, with 11 days in 2015 (**Figures 7–9**).

5.3 Blooming in Bousalem site

During the 3 years of study, "V1" and "V3" were the earliest, "V2" and "V4" triggered an intermediate blooming date, while "V5" was the last. The blooming period was spread out for all cultivars during the 2015 year and was shortened for the 2 years (2013 and 2014).

The blooming period was between 10 and 14 days during 2013 for "V1" and "V5," respectively. This period was extended during 2014 and varied from 17 to 22 days for "V2" and "V5," respectively. However, during 2015, the blooming period varied from 9 ("V4") to 16 ("V2") days. Flowering began early in 2014, late in 2015, and intermediate in 2013 for all cultivars (**Figures 7–9**).

5.4 Comparison of blooming period in the three sites

The dates and period of blooming for the 11 studied cultivars varied between the sites, cultivars of the same site, and between the years of study. The blooming periods of the different studied cultivars were superimposed on each other, which created the conditions for possible pollination between compatible or semi-compatible cultivars. Full blooming was between 6 and 16 days for all early and late cultivars in the three study sites. The cultivars of Bousalem showed a shortened blooming period during 2013 and a spread-out blooming period during 2014, explained by the difference of temperature between the years and the low chill accumulation during 2014. The Ain-Draham site is characterized by the highest chill accumulation and late blooming during the 3 years, which is explained by the effect of climatic conditions on blooming according to Westwood [14]. At each site, the blooming periods of all cultivars overlapped with each other except for the local one "Bouargoub," which was ahead during 2014 and 2015.

For this reason, the latter is not recommended as a pollinating cultivar for the others grown in Ain-Draham. According to Nyeki [34], for the sweet cherry tree, a blooming period of 10–14 days, with at least 4–6 days of full blooming, is necessary. This author mentioned that for stone fruits, a period of 3 days of overlap in full blooming is adequate, which is the case of our study in the three sites for all cultivars except for "Bouargoub" during 2015 and 2014.

In the Ain-Draham site, full blooming can vary from 5 to 16 days. Generally, it occurs during the month of April and rarely extends to the beginning of May. In Tibar, full blooming overlaps between the third week of March and the second week of April. The four cultuvars "Napoleon," "Van," "Moreau," and "Sunburst" behave differently in the two sites which can exclude the genetic potential factor in the triggering and the duration of flowering assuming that this phenomenon depends on the physiological state, age, rootstock, expression of cultivar genes, and other external factors (photoperiod, soil, nutrient supply, rainfall, and temperature).

The difference in the date and duration of blooming among the receiving cultivars (to be pollinated) and the pollinating cultivars is the cause of a fruit set failure, which is confirmed by the works of Bekefi [35], Tosun and Koyuncu [36], Beyhan and Karakaş [37], and Moghadam et al. [38]. These authors have shown that in addition to the gameto-phytic self-incompatibility (GSI), the efficiency of pollination and fertilization in the cherry tree is also affected by the availability of pollinating insects and weather conditions in particular temperature during flowering.

6. Effect of temperature (maximum) during blooming period on fruit yield

The maximum temperature during the flowering period has a negative effect on the yield at the Ain-Draham site. The cultivar "Van" is characterized by the lowest yield (3.2 Kg/tree) during 2013 and a highest temperature during blooming (17.43°C), while "Bouargoub" produced 8 Kg/tree in 2015 and bloomed during a period characterized by a low temperature (13.37°C) (**Figure 4**).

In Tibar site, the year 2015 was characterized by a low temperature during blooming and by a better yield. The cultivars "Moreau" and "Sunburst" registered 6.5 and 7 Kg/tree and a maximum blooming temperature of 20 and 22°C, respectively. While "Napoleon" records the lowest yield and a low blooming temperature (18, 19°C) (**Figure 10**).

Bousalem site, the blooming periods of the cultivars "V2" and "V5" in 2013 and 2014 were characterized by almost the same maximum temperature. The cultivars "V3" and "V1" were also characterized by the same blooming temperature, whereas they showed a difference in yield throughout the 3 studied years. The cultivar "V4" was characterized by the highest yield during the 2 years, 2014 and 2015, while its flowering period was overlapping with that of "V2" (**Figure 11**).

The temperature at the blooming period is a determining parameter for the yield. If the blooming period coincides with a low mean maximum temperature the yield is high, whereas if the blooming coincides with a high mean maximum temperature the yield will be low, which is the case for the local cultivar "Bouargoub" in the site of Ain-Draham and the cultivar "V5" in the Bousalem site. "V5" is characterized by spreading blooming and a low yield despite the highest number of stamens. However, the other cultivars such as "V1" and "V3" bloom during a period characterized by a low mean maximum temperature (19.61–17.51°C and 19.8–18.85°C, respectively) and record low yield.

These results show that the temperature during blooming determines the fruit yield in the sweet cherry tree, but there are other factors that influence

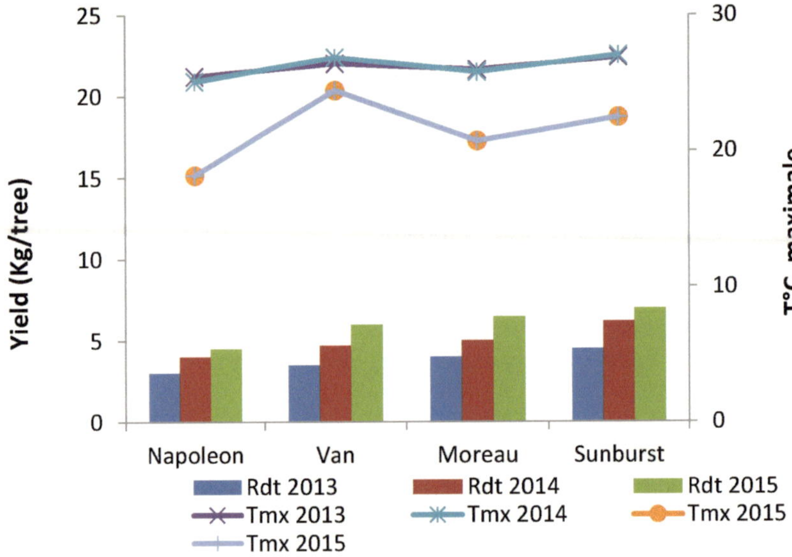

Figure 10.
Yield (Kg / tree; Rdt) and Maximum temperature during blooming at Tibar site.

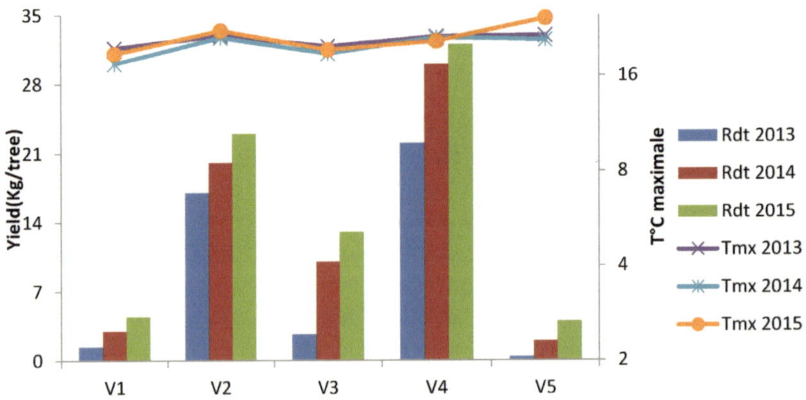

Figure 11.
Yield (Kg/tree; Rdt) and Maximum temperature during blooming at BouSalem site.

this parameter such as the genetic potential and self-fertility of the cultivar. The difference in yield between cultivars and sites can be explained by several factors including the behavior of the flower pieces depending on environmental conditions. The duration of the stigma's viability is influenced by weather factors. Regarding sour cherries, Nyeki [34] observed that the viability of the stigma was 2–3 days during sunny and hot days (mean daily temperature is 15–22°C). The viability was longer (4–6 days) in cool weather and daily temperatures of 4–12°C.

Low temperatures and rainy weather reduce the receptivity of the stigma. This was reported by Davarynejad [39] for apples and, in 1996, for pear trees. Although temperature is the main driver of phenological development, other ecogeographic factors can influence the date of flowering.

Thus, the cold temperature during blooming reduces the rate of growth of the pollen tube and can shorten the effective pollination period [40]. Caprio and Quamme [41] have shown a negative effect of high temperatures before blooming (above 27°C) on the longevity of the ovum and on the efficiency of pollination. In addition, rain and low temperatures negatively affect the activity of pollinating bees and, consequently, the setting rate [42].

7. Conclusion

Sweet cherry is sensitive to temperature profiles during the blooming period. The low productivity is largely due to the nonoverlap of flowering periods and pollen incompatibility among different cultivars in the same experimental site. Our study is based on a mixture of introduced and local cultivars with different characteristics to diversify Tunisian orchards. While, the introduction of foreigner sweet cherry cultivars in areas with mild winters leads to increased yields.

Acknowledgements

This work has been supported by the National Research Institute of Rural Engineering Water and Forests (INRGREF), Tunisia. The authors are grateful to the engineers and technicians who contributed to this work.

Conflict of interest

The authors declare no conflict of interest.

Author details

Thouraya Azizi-Gannouni* and Youssef Ammari
National Researches Institute of Rural Engineering, Water and Forests, Ariana, Tunisia

*Address all correspondence to: najeh.esak@gmail.com

References

[1] Wünsch A, Hormaza JI. Cultivar identification and genetic fingerprinting of temperate fruit tree species using DNA markers. Euphytica. 2002a;**125**:59-67

[2] F.A.O. Les engrais et leurs applications, Précis à l'usage des agents de vulgarisation agricole., Quatrième édition, Editions F.A.O., I.F.A. (Paris, France) et IMPHOS. Maroc: Casablanca; 2014. p. 84

[3] DGPA. Forest General Direction of the Agriculture Ministry; 2016

[4] Mahmood K, Carew JG, Hadley P, Battey NH. The effect of chilling andpost-chilling temperatures on growth and flowering of sweet cherry (*Prunus avium* L.). The Journal of Horticultural Science and Biotechnology. 2000;**75**:598-601

[5] Gannouni TA, Campoy JA, Quero-García J, Barreneche T, Albouchi A, Ammari Y. S-locus diversity of sweet cherry cultivars. Tunisia Journal of new sciences Agriculture and Biotechnology. 2018;**58**(1):3738-3742

[6] Schuster M. Incompatibility (S-) genotypes of sweet cherry cultivars (*Prunus avium* L.). Scientia Horticulturae. 2012;**148**:59-73

[7] Lawson EJR, Poethig RS. Shoot development in plants—time for a change. Trends in Genetics. 1995;**11**(7):263-268

[8] Castède S. Etude et compréhension du déterminisme génétique et moléculaire de la floraison chez le cerisier (*Prunus avium*) en vue de son adaptation aux futures conditions climatiques. Thèse Doctorat. Univ. Bordeaux; 2014. p. 284

[9] Balasubramanian S, Sureshkumar S, Lempe J, Weigel D. Potent induction of *Arabidopsis thaliana* flowering by elevated growth temperature. PLoS Genetics. 2006;**2**(7):980-989

[10] Éric R, Mohamed S, Abdellah L. Gestion durable des eaux et des sols au Maroc Valorisation des techniques traditionnelles méditerranéennes. 2010. IRD, ISBN: 978-2-7099-1683-7

[11] Yu H, Xia NH, Ye HG. Validation of *Prunus hainanensis* (Rosaceae). Nordic Journal of Botany. 2007;**25**:31-32

[12] Spencer R, Barley A, Pearson S. Horticultural Flora of South-Eastern Australia Volume 3: The Identification of Garden and Cultivated Plants. Sydney, Australia: University of New South Wales Press; 1995

[13] Faust M. Physiology of Temperate Zone Fruit Trees. New York: Wiley Intersciences; 1989. pp. 106-227

[14] Westwood MN. Temperate Zone Pomology. 3rd ed. Portland, Oregon, USA: Timber Press; 1993. p. 535

[15] Azizi-Gannouni T, Sghaier T, Ammari Y. Behavior and morphometric characterization of local and introduced cultivars of sweet cherries (*Prunus avium*), tested in a multi-site trial in Tunisia. Scientia Horticulturae. 2020, 2020;**270**:109455

[16] Molina-Montenegro MA, Cavieres LA. Effect of density and flower size on the reproductive success of *Northoscordum graminum* (alliaceae). Gayana Botany. 2006;**63**:93-98

[17] Johnson RS, Uriu K. Peach, Plum and Nectarine: Growing and Handling for Fresh Market. Oakland, CA: University of California, Division of Agriculture Resource; 1989. pp. 68-81

[18] Wetzstein HY, Yi W, Porter JA. Flower position and size impact ovule

number per flower, fruitset, and fruit size in pomegranate. Journal of American Society of Horticultural Science. 2013;**138**:159-166

[19] Niu G, Heins RD, Cameron AC, Carlson WH. Day and night temperatures, daily light integral, and CO_2 enrichment affect growth and flower development of *Campanula carpatica* 'blue clips'. Scientia Horticulturae. 2001b;**87**:93-105

[20] Lu Z. Temperature and genotype influence sweet cherry pollination biology [Thèse de doctorat] Université de Washinton; 2014

[21] Ouzeau G, Déqué M, Jouini M, Planton S, Vautard R, Jouzel J. Scénarios régionalisés: édition 2014 pour la métropole et les régions d'outre-mer. Le climat de la France au XXIe siècle: Ministère de l'Écologie, du Développement durable et de l'Énergie. 2014;**4**:62

[22] IPCC. Fifth Assessment Report of the Intergovernmental Panel on Climate Change. Geneva, Switzerland: IPCC; 2014. p. 151

[23] Menzel A, Fabian P. Growing season extended in Europe. Nature. 1999;**397**:695

[24] Guédon Y, Legave J-M. Analysing the time-course variation of apple and pear tree dates of flowering stages in global warming context. Ecological Modelling. 2008;**219**:189-199

[25] Chmielewski FM, Rötzer T. Response of tree phenology to climate change across Europe. Agricultural and Forest Meteorology. 2001;**108**:101-112

[26] Ricard Mp. Vérification de modèles phénologiques durant la dormmance des bourgeons de pommier et analyse de l'expression de gènes possiblement reliés à la dormance. Exigence Partielle de la Maitrise en Biologie. 2014:94

[27] Cesaraccio C, Spano D, Snyder RL, Duce P. Chilling and forcing models topredict bud-burst of crop and forest species. Agricultural and Forest Meteorology. 2004;**126**:1-13

[28] Jimenez S, Reighard GL, Bielenberg DG. Gene expression of DAM5 and DAM6 is suppressed by chilling temperatures and inversely correlated with bud break rate. Plant Molecular Biology. 2010;**73**(1-2):157-167. DOI: 10.1007/s11103-010-9608-5. PMID: ISI:000276440900014

[29] Alburquerque N, García-Montiel F, Carrillo A, Burgos L. Chilling and heat requirements of sweet cherry cultivars and the relationship between altitude and the probability of satisfying the chill requirements. Environmental and Experimental Botany. 2008;**64**:162-170

[30] Campoy JA, Ruiz D, Allderman L, et al. The fulfilment of chilling requirements and the 359 adaptation of apricot (*Prunus armeniaca* L.) in warm winter climates: An approach in 360 Murcia (Spain) and the Western Cape (South Africa). European Journal of Agronomy. 2012;**37**:43-55. DOI: 10.1016/j.eja.2011.10.004

[31] Gannouni TA, Campoy JA, Quero-García J, Barreneche T, Arif A, Albouchi A, et al. Dormancy related traits and adaptation of sweet cherry in Northern Africa: A case of study in two Tunisian areas. Scientia Horticulturae. 2017;**219**:272-279

[32] Meier U, Graf H, Hack H, Hess M, Kennel W, Klose R, et al. Phenological growth stages of pome fruits (*Malus domestica* Borkh. and *Pyruscommunis* L.), stone fruits (*Prunus* species), currants (*Ribes* species) and strawberry (*Fragariax ananassa* Duch.). Nachrichtenblatt des Deutschen Pflanzenschutzdienstes. 1994;**46**(7):141-153

[33] Tzonev R, Yamaguchi M. Investigations on some far-east *Prunus* species: Phenology. Acta Horticulturae. 1999;**488**:239-242

[34] Nyeki JM, Popovics SL. Strategy of the sour cherry verticum in the northern great plain region Hungary (analytic study). International Journal of Horticultural Science. 2002;**11**:7-31

[35] Bekefi Z. Self-fertility studies of some sweet cherry (*Prunus avium* L.) cultivars and selections. International Journal of Horticultural Science. 2004;**10**(4):21-26

[36] Tosun F, Koyuncu F. Investigations of suitable pollinator for 0900 Ziraat sweet cherry cv.: Pollen performance tests, germination tests, germination procedures, in vitro and in vivo pollinations. Horticultural Science (Prague). 2007;**34**:47-53

[37] Beyhan N, Karakaş B. Investigation of the fertilisation biology of some sweet cherry cultivars grown in the central northern Anatolian region of Turkey. Scientia Horticulturae. 2009;**121**(3):320-326. DOI: 10.1016/j. scienta. 02.028

[38] Moghadam EG, Hosseini P, Mokhtarian A. Blooming phenology and self-incompatibility of some commercial cherry (*Prunus avium* L.) cultivars. Iranian journal of Horticultural Science. 2009;**123**(1):29-33. DOI: 10.1016/j. scienta

[39] Davarynejad GH. Flowering Biology Fertility and Combination of Apple Cultivars [Ph.D dissertation]. Hung. Acad. of Sciences; 1992

[40] Sanzol J, Herrero M. The 'effective pollination period' in druit trees. Scientia Horticulturae. 2001;**90**:1-17

[41] Caprio JM, Quamme HA. Weather conditions associated with apple production in the Okanagan Valley of British Columbia. Canadian Journal of Plant Science. 1999;**79**:129-137

[42] Roversi A, Ughini V. Influence of weather conditions of the flowering period on sweet cherry fruit set. Acta Horticulturae. 1996;**410**:427-433

Section 3

Fruit Production

Chapter 5

Production Technology of Peach, Plum and Apricot in India

Bhende Siddhesh Shamrao

Abstract

In this chapter production technology of peach, plum and apricot in India is elaborated in detail in relation to introduction, origin and distribution of crop, importance and uses, morphological features of tress, other related species involved, climate and soil requirement, varieties, propagation and raising of rootstocks, planting and planting densities, cropping systems, manure and fertilisers application, cultural practices, weed management, orchard floor management, after care training and pruning, pollination and pollinizers, flowering and crop regulation, use of growth regulators, fruiting in the crop, fruit thinning and drop, maturity and harvesting, post-harvest management, handling and storage, insects, pests and diseases, special production problems like low productivity, unfruitfulness and self-incompatibility, premature leaf fall, replant problem, alternate bearing and remedies and physiological disorders of the crop.

Keywords: peach, plum, apricot, production technology, India

1. Peach

Scientific name: *Prunus persica* Family: Rosaceae.

Peach has a good position among stone fruits and is rated as the third most important temperate fruit in India. With the introduction of chilling varieties, the crop is becoming popular in the subtropical plains of North India.

1.1 Origin and distribution

Early writers were of the opinion that peach is a native of Persia. The cultivated form of peach has come from China. The scientist from Spain introduced the Peach in America, and 50 years after Cortez, it was introduced in Mexico. Peach is grown as a commercial and home fruit in most of the temperate countries of the globe. The major peach-producing countries are Italy, the USA, Spain, China, France, Greece, Russia, Mexico, Japan and Argentina. In India, it is mainly cultivated in J&K, Himachal Pradesh, Uttaranchal, Punjab and Delhi.

1.2 Importance and uses

Peach is a delicious, juicy and highly palatable fruit. It is a rich source of vitamin A, iron and protein. The fruit is generally consumed fresh, but delicious products like squashed and dried products, frozen preserves, jam, nectar, juice, beverage, marmalade, etc. can be made from it. Peaches are also a good source of low-calorie diet.

1.3 Morphological characters of plant

Peach is a small- to medium-sized upright spreading, open topped deciduous tree. The trunk bark is dark brown and rough, and young shoots are smooth and pinkish in colour. The leaves are simple, large, oblong lanceolate, glabrous above and pubescent beneath. Vegetative and flower buds are borne in the axil of the leaves. Its flowers are numerous, sessile, white or pink appearing before the leaves. The flower is of perigynous type as the perianth surrounds the pistil but is not fused to it. The floral configuration is five sepals, five petals, thirty stamens and a single ovary. Fruits are fuzzy with free- or clingstone; however, the nectarines are fuzzless peaches. The stone is deep-pitted and very hard. There are two well marked horticultural forms [1]. These include:

Clingstones: *Persica vulgaris*, Risso.
Freestones: *Persica domestica*, Risso.

1.4 Ornamental forms of peach

a. *Prunus persica var. nucipersica*, Schneid: the nectarine.

Usually small, smooth-skinned fruit, and leaves strongly serrate but not always so.

b. *Prunus persica* var. platycarpa, Bailey: saucer peach.

Flattened, scarcely thick greenish fruit with a red cheek, medium in size with good flavour.

1.5 Other species

1. Prunus davidiana

Slender, willow-like tree.
Smaller than common peach.

2. Prunus *mira.*

Small, bushy tree.
Late blooming habit.

1.6 Flowering and fruiting

The flowers are perfect, solitary, sessile and pink coloured. Flowering starts in the first week of February and continues till the end of the month. Pollination is aided by insects, and the mode of pollination is homogamy. The pollen of peach is highly viable. Commercial peach varieties are self-fruitful and set good crops without cross-pollination. J.L. Hale is the only variety which is self-unfruitful and requires to be pollinated by other varieties. Fruit setting starts in the beginning of March. The fruits are borne after 1-year growth. A small proportion is borne on short-lived spurs also. The fruit is drupe, and the edible portion is the mesocarp [2].

1.7 Climate

In India, peaches are mainly grown in midhills at a height ranging from 1000 to 1600 m. They also do well in wet and humid climate with cold winter and dry summer. It requires a chilling period below 7°C for breaking dormancy and flowering. The chilling requirement varies from 200 to 850 hours.

1.8 Soil

Peach thrives well on light sandy soils. Deep fertile loam or sandy loam with good drainage is considered to be the best. The pH ranges between 5.8 and 6.8. It cannot tolerate imperfect drainage. Fertile and heavy soils are hazardous.

1.9 Planting and aftercare

Peaches are planted in the end of January in 1 m^3 pits at a distance of 6.5 m apart. In hills, the spacing adopted is 5 × 4 m. A spacing of 3 × 3 m is practised under high-density orchards. Immediately after planting, staking is provided. The support should be firm, preferably of bamboo or other wooden logs. Young plants should be watered at frequent intervals. Stock sprouts should be removed carefully.

1.10 Propagation

The peach is commercially propagated by means of budding and grafting. T-budding is the common method, and the season of budding is from April to September. Tongue grafting and wedge grafting are also done. Grafting is done in December–January.

1.11 Raising of rootstocks

Rootstock for peaches is raised from seeds of desi peach trees. The seeds of commercial cultivars like Sharbati and Khurmani are also used. Peach plants can also be raised on peach-almond hybrid, apricot, almond seedling, plum and behmi (*Prunus mira*). Most commonly used rootstock are the seedlings of wild peach. Peach seeds need chilling in hours to germinate. The process of meeting the cold requirement is called stratification. The stratification of peach seeds can be done under natural conditions and cold storage conditions at 10°C or below.

1.12 Grafting operations

Peach seedlings from stratified seeds planted in the nursery beds during January become buddable in June. The grafting can be done when the seedlings are still in the nursery beds, or these can be uprooted and brought at one place for grafting.

1.13 Training

The peach is trained to the modified leader system. Plants should be headed back to a height of 90–100 cm at the time of planting. All the branches on the plants are also cut back to two buds. Three to five laterals which are well spaced are allowed to develop around the trunk. The lowest branch should not be below 40–50 cm from the ground level.

1.14 Open-centre system

After planting, the plant is cut back to 40–60 cm above the ground level. During the growing season, about three to six laterals, in addition to the central leader, are produced on the tree. In the first winter pruning, three to four scaffold branches which are well located and have wide angle should be selected, and the remaining unwanted branches are removed. The central leader is also completely removed. The selected branches are headed back to ¼ to ½ of the growth. During the second dormant pruning, two to three secondary branches are selected on the primary branches. The major consideration in selecting secondary branches should be their location so that after pruning, the tips of the primary and secondary leaders are about 30–40 cm apart from each other. The height of secondary branches is staggered in different years by pruning all branch leaders more severely. The vertical ones are pruned more severely. This will produce branch leaders at different heights and prevent overcrowding when the tree is mature. In the following years, the head should be fully formed, and selection of secondary branches is completed.

1.15 Tatura trellis system

In high-density planting, this system of training of plants is very popular due to yield efficiency. Trees are planted at a spacing of 5 × 1 m or 6 × 1 m. At the time of planting, a 1-year-old plant is headed back to 20 cm above the ground level. In the next growing season, two limbs or branches are selected in opposite directions, and these branches are trained across the interrow space at an angle of 60° from the horizontal, forming a V-shaped canopy. The canopy is supported by a permanent trellis constructed of high-tensile galvanised steel fence posts. The secondary branches are developed along each primary branch forming a fruiting canopy.

1.16 Pruning

The peach fruits are borne after 1-year growth. A small proportion is borne on short-lived spurs also. Pruning should be done so as to produce 50–100 cm of growth in young trees and 30–70 cm in old trees annually. About 40% of 1-year-old branches should be thinned out to ensure proper tree growth and improve fruit size and quality. The pruning of peaches is carried out in early January.

The main objective of pruning is to maintain balance between vegetative growth and fruiting. Bearing peach requires heavy and regular pruning because it bears fruits laterally on the previous season growth. It is known that once the tree bears fruits, it will never bear again throughout its life. Therefore, pruning is done to remove the unproductive parts which in turn will form new fruiting branches in the following season. In peach pruning, thinning and heading back of shoots are two basic components. Pruning should be done so as to produce 30–70 cm of growth under subtropical conditions and 25–30 cm under midhills, annually, which is sufficient for optimum fruit production. For good-quality fruit production, 40–50% of thinning out and 75% heading back of shoots are suggested under midhill conditions. At the time of pruning, dead, diseased and broken branches should be pruned off.

1.17 Orchard floor management

In the initial years of plantation, the intercrops like peas, beans, tomato, cabbage, cauliflower and ginger are grown in the vacant area in between the trees but not in the basin area. Besides these, some green manuring crops like bean, peas and gram should be grown which helps in improving soil texture and nutrient status. In bearing orchard, the basin area of trees should be kept clean either by manual weeding or the use of weedicides. Sod grasses like white clover, red clover, orchard grass and rye grass are grown in the vacant area between the trees. Basins are mulched with 10-cm-thick dry grass mulch or black alkathene mulch. The mulching helps to conserve soil moisture and efficiently control the weeds in the basin area.

1.18 Thinning of fruits

Thinning is more desirable on mature trees making small annual growth than on young vigorous plants. Thinning may also be another resort when good-sized uniform fruits are needed for canning and fancy trade. Hand thinning and chemical thinning can be done. While doing hand thinning, first shake the branches in order to dislodge the fruits which are likely to drop off naturally. If there is still surplus fruit, then start thinning from top to bottom of the branches. The distance from fruit to fruit after thinning on the shoots should be between 10 and 15 cm, and the average number of leaves per fruit should be 20–25.

1.19 Manuring and fertiliser application

Particulars	Himachal Pradesh	Punjab	Uttaranchal
FYM (kg/tree)	60	25	10
N (g/tree)	500	500	300
P2O5 (g/tree)	250	120	300
K2O (g/tree)	600	500	500

1.20 Harvesting

Peach fruits are harvested quickly when ready for harvest. Fruits are picked when they are still hard as they ripen in storage or transit.

1.21 Irrigation

Irrigation is very essential for harvesting the peaches of better size and quality. A sufficient moisture in the soil before the emergence of leaves and flowers is required for proper fruit set and growth. Frequent irrigations are needed during the fruit development. Lack of irrigation, particularly, during dry and hot summer, results in fruit drop and reduced fruit size and quality. In the hills, at least two to three irrigations and, in plains, weekly irrigation should be given during the fruit development period. In general, for quality fruit production, irrigation at 80% of field capacity is recommended. Orchard soil management and weed management are done during the initial 3–4 years after planting; the intercrops like peas, beans, tomato, cabbage, zinger and *Colocasia* are grown in between the peach trees; and basin area is mulched with hay or alkathene mulch. In fully grown tees, sod grasses are grown in vacant areas, and basin area is mulched with suitable mulch materials. Weedicides like simazine and atrazine at 2.0 kg/ha, terbacil at 0.8 kg/ha as pre-emergence and paraquat at 4.0 litre/ha and glyphosate 4.32 kg/ha as post-emergence herbicide proved to be most effective to control the weeds in peach orchards.

1.22 Crop regulation

Heavy flowering and fruiting are the characteristic features of peach trees resulting in small-sized, poor-quality fruits and reduction of flowering in the subsequent season. Hence for production of quality fruits, crop regulation through thinning is essential in peach. The criteria for fruit thinning in peach are based on leaf to fruit ratio and spacing between fruits per tree. Generally 30:40 leaves per fruit is the appropriate ratio. Application of ethephon (300 ppm) at petal fall in July Elberta is recommended for optimum fruit thinning. However, in Redhaven peach, ethephon (600 ppm), 20–30 days after fruit set when the fruitlets are 20–25 cm in diameter, should be used for thinning. Hand thinning at 5–7.5 cm fruit spacing before pit hardening stage is equally effective.

1.23 Maturity, harvesting, storage and postharvest management

Harvesting of peaches at proper stage of maturity is essential as the post-harvest quality and storage life of fruits are controlled by maturity. Various indices used for judging fruit maturity used are the number of days from full bloom, calendar dates, fruit size, firmness, pit discoloration, freeness of pit and change of ground colour. Days required from flowering to maturity in different cultivars vary from 78 to 127. Early-season varieties like Flordasun takes 81 days, Alexander 86 days, mid-season July Elberta 101 days and late-season cv. Elberta takes 127 days from full bloom to harvest. Ground colour variation in conjunction with flesh firmness is one of the best maturity indices in peaches. Peach fruits do not mature uniformly, and hence several pickings are needed during harvesting. Hand picking is the standard method for harvesting fruits. The picking containers are lined with cushion materials to avoid cuts and bruises. Immediately after harvesting, fruits are stored at a cool place or marketed. Preharvest application of calcium nitrate at the rate of 1.5% increases storage life of peaches.

1.24 Storage

Peaches have a shorter storage life than most other temperate fruits. The recommended cold storage conditions are 0–0.3°C and 85–90% relative humidity. In these conditions, freestone peaches and nectarines can be kept for 2 weeks and clingstone for 4 weeks. Precooled peaches can be stored for 28–36 days. In controlled atmosphere storage containing 5% CO_2 and 1–2% O_2 at 0°C peaches can be stored up to 42 days. The peaches came into bearing after 2 years of planting in the field. The economic bearing life of peach plant is about 20–30 years. The yielding capacity increases with the age of the plant. The average yield of fully grown trees of different varieties varies from 50 to 125 kg in hills. In conventional plantation, 7–10 tones/ha and, under high density with Tatura trellis system of training, about 23 tones/ha yield have been obtained.

1.25 Pests

Peach leaf curl aphid.
Peach black aphid.
Chafer beetle.
Peach fruit fly.
Flat-headed borer.

1.26 Peach leaf curl aphid (*Brachycaudus helichrysi*)

This is the most serious pest of peach. They also infest plum plants. The aphids suck sap from the buds and sprouting foliage causing curling, yellowing and thickening of leaves. The activity of aphid is seen with the emergence of new growth during March. Floral buds also become weak which results in poor setting.

1.27 Control

The pest is controlled by spraying of 0.025% methyl demeton (200 ml metasystox 25 EC) or 0.03% dimethoate (200 ml Rogar 30 EC) in 200 litres of water 7–10 days before flowering. The spray should be repeated after 5 days.

1.28 Diseases

Shot hole.
Peach leaf curl.
Powdery mildew.
Brown rot.
Viruses.
Crown gall.

1.29 Varieties

State	Early	Mid-season	Late
Himachal Pradesh	Alton, World's Earliest, Early White Giant, Redhaven, Stark Red Gold	July Elberta, Kanto 5, Shimizu Hakuto, Sunhaven	J.H. Hale
Uttar Pradesh (midhills)	Early Candour, Redhaven, Sunhaven	July Elberta, Alexander Crawford Early	Parrot Delux, J.H. Hale Peregrine
Jammu and Kashmir	Peshwari, Quetta, July Elberta, Saharanpur Prabhat	J.H. Hale, Alexander and CO Smith	

1.29.1 Candour

Early peach of the season.
Yellow and medium–sized fruit.
Fruit tends to produce split seeds, which limits this variety's usefulness in canning.

1.29.2 Redhaven

Early variety suited to midhills.
Medium-sized round fruit.
Yellow skin with red tinge.
Firm, fine-textured, melting flesh.

1.29.3 Alexander

Excellent, early-season cultivar.
Medium- to large-sized round fruit.
Smooth, beetroot purple skin with green patches.
Soft, greenish white, juicy, sweet, aromatic flesh.
Freestone variety.

1.29.4 Elberta

Suitable for canning.
Large, round fruit.

Pale yellow skin with red splash.
Firm, juicy, sweet flesh.

1.29.5 Stark Red Gold

Early variety.
Medium-sized, moderately fuzzy fruit.
Orange skin with a blush of red.
Semi-clinging stone.
High-yielding variety.

1.29.6 Sunhaven

Medium-sized round fruit.
Bright red skin.
Yellow, firm, fine-textured flesh with good flavour.
Freestone variety.

1.29.7 Partap

Round, yellowish red fruit.
Firm, yellow flesh.
Freestone variety.
12% TSS and 0.7% acidity.

1.29.8 Sharbati

Spreading and vigorous tree.
Large, greenish yellow fruit with rosy patches.
Juicy, tasty, white flesh with excellent flavour.

1.29.9 Khurmani

Red-coloured large fruit pointed at the base.
Clingstone cultivar.
White, soft, juicy flesh.

1.29.10 Nectarine

Fuzzless peach.
Smooth skin.
Smaller in size.
Firm flesh.
Stronger flavour and aroma.

2. Plum

Scientific name: *Prunus salicina.*
Family: Rosaceae.

2.1 Introduction

It is a strong growing temperate tree. This fruit have found favour with orchardists because of their phenomenal yield potential and high economic returns. It was introduced to India by Alexander Court in his orchard at Mashobra. The main producing countries are Russia, Romania, China, Germany, the USA, France, Italy and Spain. In India, it is grown in J&K, Himachal Pradesh, Uttaranchal and Punjab. It is also grown to some extent in Nilgiri hills of South India.

2.2 Importance and uses

Besides being palatable and delicious, the fruit has high nutritive value also. It is one of the richest sources of vitamin B1. Plum fruit is also rich in vitamin A and riboflavin. It is a good source of sugars, proteins, carbohydrates and minerals like calcium, phosphorous and iron. It is also known for its cooling effect, and it is considered best to overcome jaundice.

2.3 Morphological characters

Tree is medium- to large-sized, upright growing and deciduous. Leaves are alternate, serrate, sharp-pointed, medium-sized and glabrous. Flowers are produced three in a bud on a 1-year shoot or spur. Flowers are perfect, solitary or raceme, with five sepals; five petals, usually white; numerous perigynous stamens; one pistil with elongated style; and two ovules; and the fruit is drupe and usually single seeded. White flowers are seen clustered on the spurs and come to full bloom 7–10 days after the emergence of first flower. The fruit setting starts in the second week of March. Fruit is drupe, and the edible portion is the mesocarp.

2.4 Climate and soil

Plum requires varying types of climate and is grown from subtropical plains to the temperate high hills. The European-type plums require temperate climate and are grown in high hills at an elevation of 1300–2000 m ams1. It requires about 800–1000 h of chilling below 7°C during winter to break rest period. Japanese plum requires 100–800 h of chilling, and winch is met in midhill areas located at an elevation of 1000–1600 m amsl. Plums can be grown in areas where winters are cold and summers are hot. Cold, wet and windy weather during bloom is detrimental for good fruit set as spring frost injury causes damage to bloom. A northern slope is preferred particularly for Japanese plum, which tends to delay the bloom period and thus avoids early frost injury. Plum requires 90–110 cm well-distributed rainfall throughout the year. Prolonged drought during fruit growth and development and excessive rains during fruit maturity hamper fruit quality. Although plum can grow on a wide range of soils, deep, fertile and well-drained loamy soils with a pH of 5.5–6.5 are the most suitable. The soil should be free from hard pan, waterlogging and excessive salts. Very heavy or light soils are not suitable. The Japanese plums do well on average soils having high pH.

2.5 Species and varieties

The cultivated plums belong to two species:

1. *Prunus domestica* (European plum). It is a hybrid of diploid myrobalan plum (*Prunus cerasifera*) and tetraploid black thorn (*Prunus spinosa*). It is hexaploid in shape. Fruits are larger in size than Japanese plums. Fruit is oval or round having both yellow and green ground colour and also both red and blue skin colour. The cultivated varieties of European plum are classified into three main groups:

a. Prunes. Fruit is oval in shape with bulging ventral side and compressed bilaterally. It is blue or purple in colour, high in sugar content which makes itself suitable for drying without removal of pit. All prunes are plums but all plums are not prunes. Varieties are Italian prunes, Giant prune and President.

b. Reineclaude and green gage plum. This is a hybrid between *Prunus domestica* and *Prunus insititia*. Fruit is greenish yellow in colour and round in shape having yellow skin and flesh. Important varieties are Golden Drop, Green gage and Golden transparent.

c. Lombard plum. The colour of fruit is purplish red. Varieties are Lombard and Victoria.

2. *Prunus salicina* (Japanese plum). It originated in China but introduced in Japan from where it is disseminated around the world. Plant is more vigorous, productive, precocious and resistant to diseases than European plum. The fruits are large and heart shaped with pronounced apex. A few cultivars are oblate or round.

3. *Prunus insititia.* This is a small European plum which is hexaploid in shape and grows wild in Europe and Western Asia. Plums of this species are known as Damson and Mirabelles. Fruits are small and purple (Damson) and yellow (Mirabelles). Plants are small and compact and forms excellent hedge rows.

2.6 European plum cultivar

California Blue, Washington, French Prunes, Early Italian, Stanley, Grand Duke, Victoria, Damson.

2.7 Japanese plum cultivar

Beauty, Methley, Santa Rosa, Kelsey, Mariposa, Satsuma, Burbank Red Beaut, Fronteir.

2.8 Propagation and rootstocks

Plums are generally propagated through hardwood cuttings. In case rootstocks are to be used, budding and grafting methods are employed for its propagation. Plum is raised on seedling rootstock of wild apricot (Zardalu) and *Prunus cerasifera.*

2.9 Raising of rootstocks

The rootstocks can be raised either through seeds or through cuttings.

a. Through seeds. Stones are extracted from ripe fruits and sown during November in rows. Light irrigation is provided after sowing.

b. Stem cuttings. Stem cuttings are planted in beds in row at a distance of 15 cm in rows which are 30 cm apart. The rooted cuttings are budded in the following May–June or grafted in December–January.

2.10 Rootstock

Plums are propagated vegetatively by budding and grafting on rootstocks. These can also be propagated by hardwood cutting and by leafy softwood cuttings under intermittent misting. Cuttings taken from hardwood and semi hardwood are treated with IBA (2000–5000 ppm) for better rooting. For raising the seedlings, the seeds of wild apricot are stratified under alternate layers of moist sand for 45–50 days at temperature 3–5°C to break the rest. The stratified seeds are sown in nursery beds 6–10 cm deep in row 25–30 cm apart with a distance of 10–15 cm from seed to seed. The seedlings attain graftable size in a year. In Punjab, own-rooted plants of Kala Amritsari are generally used for planting. Clonal rootstocks of plum are multiplied commercially by layering. It has been observed that most of clonal rootstocks of plum are difficult to root. Application of 2500 ppm IBA to stool shoot helps to improve rooting. Clonal rootstocks are also propagated by hardwood cuttings. The hardwood cutting should be taken during dormant season and dipped in 2500 ppm IBA solution for 30 seconds. Then these cutting are planted in mist propagation chamber for rooting.

2.11 Propagation of scion

Seedling as well as clonal rootstocks which are 0.8–1.2 cm in thickness are grafted in February with tongue and cleft method of grafting. The seedling which do not attain graftable size in February, they should be budded with T- and chip methods of budding in June–July. For grafting the scion, wood should be collected from healthy, disease-free, true to variety trees during January. The scion wood is collected from 1-year-old shoots. They are packed in moist sphagnum moss, which are properly labelled for variety. These bundles of scion wood are stored in cold storage or buried deep in the soil at shady place till grafting is done. The best time of grafting of these fruits is February in lower elevation and in March at higher elevations. Chip budding can also be done in March, July and September.

2.12 Vegetative propagation

2.12.1 Hardwood cuttings

The rooted cuttings can be directly used without budding. The cuttings taken from subapical portions planted after dipping in IBA and callusing for about a month give higher percentage of success. The time of preparation of cutting is from the end of December to the end of January.

2.12.2 Budding and grafting

Budding by shield or T-method is performed in April and July–August. The plants budded becomes saleable after 1 and a half year. The propagation by cleft and tongue grafting is performed when the stock and scion are still dormant. Cleft grafting is generally performed when the stock is thicker than the scion.

2.12.3 Planting operations

They are planted in January in 1 m^3 pits. Fill the pits with top soil and well-rotten FYM before planting. Square system of planting is adopted with 6 m spacing. Provide irrigation soon after setting plants in pits and staking. Avoid water stagnation.

2.13 Planting and planting density

Planting of plum is done in December–January when the plant is in dormant conditions. Before planting, the bushes and weeds on the site of an orchard should be properly cleared off. It is advisable to plough the plantation area. In hilly area, terraces should be kept inwards to facilitate soil conservation. The orchard area should be properly laid out about 2 months before planting. In sloppy land, layout of an orchard should be done with contour or terrace system, while in a flat land, square system is followed.

2.14 Training and pruning

Plums are generally trained on the open-centre system or to modified central leader system depending upon the varietal growth habits. The varieties with spreading habits of growth as in Japanese plum should be trained to open-centre system. In HP open-centre system is followed irrespective of variety. However, in the plains, where plenty of sunlight is available, trees should be trained in the form of modified central leader system with four to five scaffold branches.

2.15 Open-centre system

After planting, the plant is cut back to 40–60 cm above the ground level. During the growing season, about three to six laterals, in addition to the central leader, are produced on the tree. In the first winter pruning, three to five scaffold branches which are well spaced and have wide angle should be selected, and the remaining branches are removed.

The central leader is also completely removed. The planting distance varies according to the varieties, rootstocks and fertility of soils. Plums are generally planted at a distance of 6 m × 6 m. In high-density plantation, the plum plants raised on semi-dwarfing clonal rootstocks should be planted at 4 m × 4m in distance. Pits of 1 m × 1 m × 1 m size are dug in October–November. Pits should be filled with fertile top soils mixed with 40 kg of well-rotten FYM and 1 kg of single super phosphate. To avoid any damage from insects, the pits are drenched with 10 litres of chlorpyrifos (4 ml/L) solution. The graft union should be kept 10–15 cm above the ground level at the time of planting to avoid any scion rooting. Young plants should be watered regularly, and basin area is mulched with 15-cm-thick dry grass.

The selected branches are headed back to ¼ of the growth. During the second dormant pruning, two to three secondary branches are selected on the primary branches. The major consideration in selecting secondary branches should be their location so that after pruning, the tips of primary and secondary leaders are about 30–40 cm apart from each other. The height of secondary branches is staggered in different years by pruning all branch leaders more severely. The vertical ones are pruned more severely. This will produce branch leaders at different heights and prevent overcrowding when the tree is mature. In the following years, the head should be fully formed, and selection of secondary branches is completed.

2.16 Pruning

In plums, thinning and heading back of shoots are two basic components of pruning. Most of the plum varieties bear on spurs on 2-year-old woods. The life span of these spurs is 5–6 years. It is necessary to prune for some spur renewal each season. The extent of pruning is done such a way to induce an annual shoot growth of 25–50 cm. In bearing plum trees, 25–30% thinning of shoots and 50–75% heading back of shoots are suggested for proper fruiting. At the time of pruning, dead, diseased and broken branches should be pruned off.

2.17 Nutrition

Plum requires adequate amount of nutrients for better growth and quality fruits. Application of manures and fertilisers depends upon soil fertility, type of soil, topography, age of tree, cultural practices and crop load. The requirement of fertilisers varies from region to region. The farmyard manure along with full dose of P and K should be applied during December and January. Half dose of N is applied in spring before flowering and the remaining half a month later.

2.18 Irrigation

Plum is mostly grown under rainfed conditions. However, in order to produce fruits of good size and better quality, irrigation is essential. Various methods of irrigation are adopted to irrigate plum orchards, but in hill basin, drip irrigation methods are more popular and are widely used and recommended. After fruit setting, the plum trees are irrigated at weekly intervals and six to eight irrigations which are recommended for higher production of quality fruits in Santa Rosa plum.

2.19 Orchard floor management and weed management

In plum orchard, sod culture and mulching of tree basin area with hay mulch or black alkathene mulch are the most common methods of orchard floor management. During pre-bearing stage, intercropping with legumes and vegetables is also practiced in orchards planted in flat and less sloppy land. The weed removal manually is one of the practices employed in the orchards besides inter cultivation and cover crop growing. In rainy seasons, the weeds in plum orchard are controlled with the post-emergence sprays of glyphosate at the rate of 800 ml/hectare.

2.20 Crop regulation and quality improvement

Generally plum tends to bear heavy crops and bear undersized fruits of low quality, and thinning, therefore, is necessary to increase the fruit size and uniformity in colour and to stimulate flower initiation for the regulation of next year's crop. Various methods such as ,manual, mechanical and use of chemicals are used, but chemical thinning has superiority with respect to thinning cost, fruit size and quality. NAA at 20 to 40 ppm sprayed after petal fall resulted in good fruit thinning.

2.21 Maturity, harvesting and yield

It has been observed that plum usually ripen unevenly over the tree. Fruits, therefore, are harvested in two or three pickings, and it is very important to find out the exact stage of picking when they are mature. Among various indices of maturity, most commonly used are flesh firmness 5.9 ± 0.45 kg, days from full bloom (94 ± 3), TSS 13.5–14.5 and TSS acidity ratio 1.2:1.5. The fruits are harvested with stalk intact avoiding any skin injury. Fruits are very delicate and perishable; therefore picking baskets should be lined with soft material on the inner surface. Immediately after plucking the fruits should be kept under the shade of the tree to remove field heat.

2.22 Grading and packing

To obtain high price in the market, the grading of the fruits is done to have uniform size and better quality. The packing and grading standards of the plum are:

2.23 Storage and marketing

Plums being perishable have very short shelf life. In India the work in HP has revealed that plum can be stored for 1 to 2 weeks at 0°C with 80–90% humidity. The CA storage has been practiced overseas by maintaining 2–3% oxygen and 2–8% CO_2, and the fruits can be retained for a duration of 2–3 months.

2.24 Flowering and fruiting

Flowering starts in the second fortnight of February and last up to the first week of March. Growth pattern follows a double sigmoid curve. Specific gravity decreases from fruit set till maturity.

2.25 Manuring and fertilisation

State	FYM (kg/tree)	N (g/tree)	P_2O_5 (g/tree)	K_2O (g/tree)
Punjab	36	180	90	216
Himachal Pradesh	40	500	250	700

2.26 Training and pruning

2.26.1 Training

The tree is trained to modified leader system and headed back to a height of 90 cm from ground level. Select four to five well-spaced laterals to grow.

2.26.2 Pruning

Annual pruning is done in January. Remove thin and crowding twigs and branches. Thin out criss-crossed, lengthy, dried and diseased branches. After every 4–5 years of fruiting, heavy pruning should be done by heading back lengthy branches about half of their length.

2.26.3 Thinning

Hand thinning is commonly practised. First shake the branches to dislodge fruits which are likely to drop off naturally. If still surplus is there, then start thinning from top to bottom. Hold the stem of fruit to be removed between the thumb and the second finger, and pull it off gently. Foliar spray of 200 ppm ethephon and 100 ppm carbaryl at full bloom is very effective for blossom thinning.

2.26.4 Harvesting

It is a climacteric fruit. So it should be picked at proper stage of maturity. It starts yielding within 2–3 years of planting. Peak season is May.

2.27 Pests

San Jose scale.
Plum twig borer.
Plum fruit moth.
Mites.
Nematodes.

2.28 Diseases

Bacterial canker.
Brown rot.
Crown rot.
Cytospora canker.

2.29 Varieties

2.29.1 Santa Rosa

Upright trees.
Large purple crimson fruit.
Amber coloured flesh.
Prolific bearer.
Very juicy fruit.

2.29.2 Red Beauty

Spreading tree.
Medium-sized fruit; mix of red and yellow colour.
Yellow-coloured flesh.
Sterile pollens; require pollinizer.
Early maturing variety.

2.29.3 Mariposa

Upright growing tree.
Requires cross-pollination.
Heart-shaped fruit.
Skin: greenish yellow mottled with red.
Red-coloured flesh.
Good shelf life.

2.29.4 Frontier

Vigorous, upright, high yielding.
Partially self-unfruitful.
Red-purple-coloured large fruits.
Red, firm, sweet flesh.
Santa Rosa when used as pollinizer improves the yield.

2.29.5 Satluj Purple

Bright crimson fruit; thick-skinned.
Sweet taste.

Yellow-orange firm flesh.
13–14% TSS, 0.6–0.7% acidity.
Self-unfruitful; requires Kala Amritsari as pollinizer.

2.29.6 Kala Amritsari

Self-fruitful.
Low spreading vigorous tree.
Medium-sized, round fruits; dark brown coloured.
Yellow flesh, moderately juicy.
15% TSS, 1.2% acidity.
Yield will improve if pollinated with Titron.

2.29.7 Titron

Self-fruitful.
Improved yield, if Alucha Early Round is used as pollinizer.
Spreading tree.
Small- and medium-sized fruit; deep purple colour, thin skin.
Yellow flesh, moderately juicy.
Excellent for jam making.

2.29.8 Alu Bhokra

Upright, tall, vigorous tree.
Self-unfruitful.
'Howe' is a good pollinizer.
Large fruit, red coloured.
Sweet and juicy pulp.

3. Apricot

Scientific name: *Prunus armeniaca.*
Family: Rosaceae.

3.1 Origin and history

The apricot ranks second next to plum among the stone fruits in India in area, production and popularity. It is a drought-resistant, salt-tolerant, hardy plant being rather less susceptible to pests and diseases. The word 'Armeniaca' indicates its introduction to Italy and Greece by Armenian traders. The major producing countries are China, Russia, Turkey, Italy, Spain, France, Greece, the USA, Morocco, Syria and Romania. In India it is grown in Jammu & Kashmir, Himachal Pradesh, Uttaranchal and limited extent to Amritsar and Patila in Punjab.

3.2 Botany

It is a small tree growing up to 6–9 m in height. The bark is reddish with glabrous twigs. The leaves are ovate, glabrous above and pubescent beneath. Pinkish to white flowers are produced before the leaves. Fruit is velvety when young and peach-like in colour and shape, and flesh is yellowish orange. Stone is smooth and flattened [3].

3.3 Climate

Apricot can be successfully grown at an altitude between 900 and 2000 m amsl. The white-fleshed sweet kernelled apricots require cooler climate and are grown in dry temperate regions up to 3000 m amsl. The long cool winter (300–900 chilling hours), frost free and warm spring are favourable for fruiting. Average summer temperature between 16.6 and 32.2°C is suitable for growth and quality fruit production. In general, the sites located on the northeastern aspect at lower elevation and southwestern at higher elevation are suitable for its cultivation. Spring frost causes extensive damage to the blossoms, which are killed when temperature falls below 4°C. An annual rainfall of about 100 mm, well distributed throughout the season, is good for normal growth and fruiting spur favours fruiting. An annual rainfall of 100 cm is sufficient for obtaining a good apricot crop.

3.4 Soil

Apricots are quite hardy and can be grown in most of the soils. Deep well-drained soil is the best. The soil should be about 3 m deep. In Kinnaur region of Himachal Pradesh, where apricots are grown in large wild stands, the soils are sandy and well-drained but not very fertile. If the drainage is good, high lime content of the soil does not depress the growth of tree.

3.5 Propagation

Apricots can be commercially raised by vegetative methods of propagation like budding and grafting. Seed propagation can also be practised.

3.5.1 Seed propagation

Apricot seeds require stratification for 72 days at 4°C. The germination of seeds can be hastened by the removal of kernel from the shell, scarification and gibberellic acid or kinetin treatment. Seeds are soaked in 500 ppm GA3 or 5 ppm kinetin solution for 24 hours before planting. Seeds are sown in nursery beds 30 cm apart. Frequent watering and mulching should be given.

3.5.2 Vegetative propagation

Apricot seedlings are generally propagated on wild apricot, wild plum and peach. Apricot seedlings are the best for graft union and for producing vigorous trees. In rainfed orchards, where drought conditions prevail, apricot on peach makes better growth than on apricot seedlings. T-budding and tongue grafting are adopted.

3.6 Planting

The apricot plants are planted during the dormant season, i.e. end of December to February, but early planting gives better establishment of plants. Undesirable trees and shrubs should be removed from the land during its initial preparations by digging and ploughing. On the flat land, a regular planting layout system such as square and triangular is followed, while on the hill slopes, contour system is generally practiced. The spacing of plants varies with the soil, climate and vigour of variety and rootstocks. The plants are generally planted at a spacing of 6 m × 6 m. The pits of 1 × 1 × 1 m dimension are dug about a month before planting and are filled with a mixture of soil and 50–60 kg well-decomposed FYM. About 1 kg SSP and 10 L of chlorphyrifos (4 ml/1 litre of water) is also added to each pit. In comparison to other temperate fruits, high-density planting in apricot has begun rather late as there are very less dwarfing rootstocks. A density of 7200 trees/ha has been reported in cv. Canino.

3.7 Flowering and fruiting

In apricot, usually three flower buds develop in the axil of a leaf at each node on a shoot and a spur. The central one is vegetative and side ones are floral. In Himachal Pradesh, flowering occurs in March. High temperature, low humidity and wind shorten flowering duration by increasing respiration.

3.8 Irrigation

Tree has shallow roots; so good soil moisture is beneficial. The roots are confined to 2 m from the surface, so wet the land up to a depth of 2 m. In rainfed conditions, mulching and water harvesting technique can be practised.

3.9 Manuring and fertilisation

Parameter	N	P_2O_5	K_2O
General	100–200 kg/ha	80 kg/ha	100 kg/ha
Kashmir	450 g/tree	150 g/tree	750 g/tree

3.10 Training

Apricot is trained to modified leader system. The main branch is cut 50–75 cm above the ground. Four to six well-spaced branches are retained.

3.11 Pruning

3.11.1 Pruning of young trees

The main central axis is cut off 50–75 cm above the ground, and all laterals are cut off if these are not properly spaced along the central axis.

Pruning at the end of first growing season in winter consists in selection of five to six well-spaced laterals around the trunk. The lower branch should be about 40–45 cm, and all other laterals should be completely thinned out. All the selected laterals should be headed back to get secondary branches on them.

During the second year growth, only five to seven secondary scaffolds about breast height are retained, and others are removed.

At the end of the third year, pruning is confined to thinning of the branches which are either crossing or crowding each other for proper development of the framework and to admit sunlight in the tree centre to promote growth of the spurs.

The leader of the tree is modified after the fourth year of its age by cutting it back very close to a lateral branch.

3.11.2 Pruning of bearing trees

To produce new spurs annually and to replace the older and unproductive ones, light to moderate thinning of branches and heading back of new laterals are essential annually. In old trees when the growth becomes less, the heavier pruning is done by way of cutting back the main primary limbs and thinning of undesirable secondary laterals.

3.12 Thinning of fruits

The apricot tends to set heavily. So to produce fruits of marketable size, it must be thinned. Fruit-to-fruit spacing should be kept 4–8 cm when crop is heavy, and when crop is light, spacing may be done to 4–5 cm. Two to three fruits may be retained on each spur. Thinning should be done within 40 days after full bloom.

3.13 Flowering

In apricot, usually three buds develop in the axil of a leaf at each node on a shoot and spur. The central one is a vegetative bud, and the two side buds are floral. Time of flowering and its duration vary with the variety and the prevailing weather conditions. Under midhill condition, the flowering in apricot occurs in the month of March and higher hills at the end of March and April.

3.14 Pollination and fruit set

Most of the commercial cultivars of the Apricot are self fruitful and set fruits without pollinizer. However, varieties like Charmagz and Perfection have been reported self-incompatible. There is generally a good fruit set in the apricot covariants growing in appropriate climatic conditions. There is 40–60% fruit set in the cultivars commercially grown in midhills, but fruit drop is to the extent of 79% in these cultivars, which occurs mostly in the second week after fruit set. A spray of 10 ppm NAA at the beginning of pit hardening reduced the preharvest drop.

3.15 Manuring and fertilisers

Apricot trees remove large amount of nutrients from the soil and require organic organic manures as well as chemical fertilisers for normal growth and fruit production. The manurial requirements depend upon age of tree, type of soil, climate conditions and cultural practices, which vary from region to region. FYM is applied during December–January along with full dose of P and K by broadcasting method. Nitrogen is applied in two doses via first half dose of N in spring 2–3 weeks before flowering and the remaining half N a month later, if irrigation facilities are available. Under rainfed conditions the second half dose of N should be applied at the onset of monsoon rains or through one or two foliar sprays of 0.05% urea after fruit set. Fertilisers should be broadcast on the soil surface under the spread of the trees and mixed with the soil. It should not be applied in too wet or too dry soil. In high rainfall areas with steep slopes, the band application of nitrogenous fertilisers should be preferred over broadcasting. During the initial 3–4 years of orchards' life, when the plants are young, intercropping with leguminous crops like pea, bean, soya bean, cowpea and also tomato and strawberry is recommended as they enrich the soil and also give economic returns.

3.16 Irrigation

Though apricot is tolerant to dry atmosphere yet requires irrigation water especially during critical periods of fruit growth and development. Water requirement varies with the soil, tree age, climatic conditions and irrigation method. The peak water use period is from the end of April to mid of June, which coincides with fruit development period. Irrigation at 20% depletion of soil moisture from field capacity improves fruit size and yield. Irrigation interval should be 10 days during May and 6–8 days during June. All in all, eight irrigations in a season are sufficient for apricot in midhill of Himachal Pradesh.

3.17 Maturity indices, harvesting and yield

Change of surface colour, days from full bloom to harvest and fruit TSS are considered to be the best indices of maturity. For fresh market, the fruits are plucked when surface colour turns green to yellow. Fully ripe fruits are, however, harvested for freezing, canning and drying. In Himachal Pradesh, days from full bloom to harvest and fruit TSS have been standardised for different varieties to judge the optimum time of harvest. Since apricot fruits are very perishable, so due care is required during harvesting, packing and transportation. The fruits should be harvested in the morning hours, and direct exposure of fruits to the sun is avoided during grading and packing. Apricot trees start fruiting at the age of 5 years and give economic yield up to 30–35 years. Apricot attains full bearing age at about 8–10 years and yield about 50–80 kg fruits per trees.

3.18 Thinning

Fruit set in apricot is rather heavy which results into undersized fruits and also increases the tendency of biennial bearing. Fruit thinning improves fruit size, promotes regular bearing, decreases limb breakage due to heavy crop load and maintains the tree vigour. Fruit thinning should be done within 40 days after full bloom, i.e. during the last week of April or first week of May, because this is the effective period influencing fruit bud formation. Both hand and chemical thinning method are employed. Depending upon the crop load, the fruit may be thinned till the fruit are 6–10 cm apart. A spur should have not more than two fruits. Foliar spray of 25–50 ppm NAA 20 days after fruit set is best for thinning.

3.19 Orchard soil management and weed management

Sod culture plus mulching of basin with dry grass or black polythene is the common orchard floor management practice followed in apricot orchards. In apricot orchards atrazine or diuron at 4.0 kg/ha as pre-emergence and gramoxone at 2 kg/L or glyphosate at 800 ml/ha as post-emergence has been found effective and economical in controlling weeds. Mulching of trees basin with 10–15-cm-thick dry grass also checks the weed growth.

3.20 Maturity indices

Fruit and flesh colour.
Size, firmness.
TSS, acidity.
Dry matter content.

3.21 Pests

Indian gypsy moth (*Lymantria obfuscata*).
Leaf roller (*Archips micaceana*).
Peach stem borer (*Sphenoptera lafertei*).
Peach fruit fly (*Bactrocera dorsalis*).

3.22 Diseases

Bacterial canker and gummosis (*Pseudomonas syringae*).
Powdery mildew (*Sphaerotheca pannosa*).
Brown rot (*Monilinia laxa*, *M. fructigena*, *Sclerotina*).
Wilt (*Verticillium albo-atrum*).

Author details

Bhende Siddhesh Shamrao
College of Horticulture, Hiriyur University of Agricultural and Horticulture Sciences, Shivamogga, Karnataka State, India

*Address all correspondence to: siddhort@gmail.com

References

[1] Bal JS. Fruit Growing. Ludhiana, Punjab: Kalyani Publishers; 2006

[2] Chadha KL, Pareek OP. Advances in Horticulture. Vol. 4. New Delhi: Malhotra Publishing House; 1993

[3] Chadha TR. Text Book of Temperate Fruits. New Delhi: ICAR; 2001

Chapter 6

Sea Almond as a Promising Feedstock for Green Diesel: Statistical Optimization and Power Rate Law Based Chemical Kinetics of Its Consecutive Irreversible Methanolysis

Chizoo Esonye, Okechukwu Donminic Onukwuli, Akuzuo Uwaoma Ofoefule, Cyril Sunday Ume and Nkiruka Jacintha Ogbodo

Abstract

For successful industrial scale-up and effective cost analysis of transesterification process, presentation of complimentary research data from process optimization using statistical design techniques, chemical kinetics and thermodynamics are essential. Full factorial central composite design (FFCCD) was applied for the statistical optimization of base methanolysis of sea almond (*Terminalia catappa)* seed oil using response surface methodology (RSM) coupled with desirability function analysis on quadratic model. Reaction time had the most significant impact on the biodiesel yield. Optimum conditions for biodiesel yield of 93.09 wt% validated at 92.58 wt% were 50.03°C, 2.04 wt% catalyst concentration, 58.5 min and 4.66 methanol/oil molar ratio with overall desirability of 1.00. Ascertained fuel properties of the FAME were in compliance with international limits. GC–MS, FTIR and NMR characterizations confirmed unsaturation and good cold-flow qualities of the biodiesel. Based on power rate law, second-order kinetic model out-performed first-order kinetic model. Rate constants of the triglyceride (TG), diglycerides (DG) and monoglycerides (MG) hydrolysis were in the range of 0.00838–0.0409 wt%/min while activation energies were 12.76, 15.83 and 22.43 kcal/mol respectively. TG hydrolysis to DG was the rate determining step. The optimal conditions have minimal error and would serve as a springboard for industrial scale-up of biodiesel production from *T. catappa* seed oil.

Keywords: kinetics, methanolysis, optimization, response surface methodology, sea almond

1. Introduction

The application of biodiesel as an alternative energy source to petrodiesel due to its various established renewable advantages has been reported by many researches [1].

Most importantly the biodiesel production from low cost feedstocks (mostly from agro-waste) that are readily and widely available, with high oil yield, non-food competing and underutilized are key parameters that make them satisfactory to EU sustainable biofuel directives. Various methods have been established as ways of converting vegetable oils into petrodiesel-replaceable-form for application in diesel engines (DE). It is very important to highlight that pyrolysis as thermal degradation of vegetable oil produces more bio-gasoline than biodiesel fuel, micro-emulsion results have been only on short term while dilution of vegetable oil with petrodiesel requires very low concentration of vegetable oil. Transesterification is therefore the major chemical process that involves the conversion of fatty acids or triglycerides in vegetable oils to biodiesel (alkyl esters). Structures of chemical building blocks (CBB) involved in transesterification process are presented in **Figure 1**. Although transesterification of vegetable oils can be conducted with both homogeneous (acid or base) and heterogeneous catalysts, base methanolysis always provide much faster rates [2] and cheaper process [1] and more predominantly applied for industrial purposes and large scale biodiesel production. However, currently the major challenge of biodiesel application as a replacement to petrodiesel is its industrial production sustainability. This can be achieved through detailed established transesterification viability and most importantly feasibility data.

Consequently, the successful scale-up of laboratory results in transesterification requires so much information obtained through optimization and kinetics studies. Hence, for effective cost analysis of transesterification process, a holistic presentation of research data from process optimization using statistical design techniques and knowledge of chemical kinetics are essential in establishing the optimum conditions, feed compositions, degree of conversion and recycling as well as reaction mechanisms. It has been established that one factor at a time (O-F-AT) has obvious challenges of non-reliability of obtained results, non-depiction of the interactive effects of the independent variables and ineffectiveness due to the existence of multiple experimental run [3]. Therefore, many researches on optimization and modeling of transesterification process have been established through the application of such soft computing techniques like response surface methodology (RSM), artificial neural network (ANN) and integrated models (IM) [4]. It is very interesting to write that RSM has undisputable edge over others due to its ability to navigate the design space, flexibility, robustness in establishing the optimum condition with the help of desirability function and capability to minimize the number of experimental runs needed to give adequate evidence for statistically acceptable result [4]. Other obvious advantages are its availability in most statistical software, as an asset in statistical quality control, expression and inferential statistics, reliability, gage repeatability and reproducibility studies and process ability as well as improved grappling output. The integration of RSM with desirability function has been reported to have a high a potential over conventional RSM [5]. In the RSM design of experiment, different types have been applied such as full factorial, fractional factorial, Box-Behken, Placket-Burman, central composite rotatable design (CCRD) etc. However, full factorial central composite design (FFCCD) has the advantage of providing double factor axial points at a fixed distance from the centre and significant replicate points at the centre. This has a resultant effect of providing a better reduced-cost approach in obtaining optimal response with least number of experimental runs.

Also, it has been reported that lack of the vital kinetics data of many non-conventional biodiesel feedstocks possess great challenges on their industrial scale process application, reactor design, simulation and control [1]. Although, many researchers have previously reported the kinetics of base-catalyzed transesterification of conventional feedstocks, those works have dwelled more on the reversible

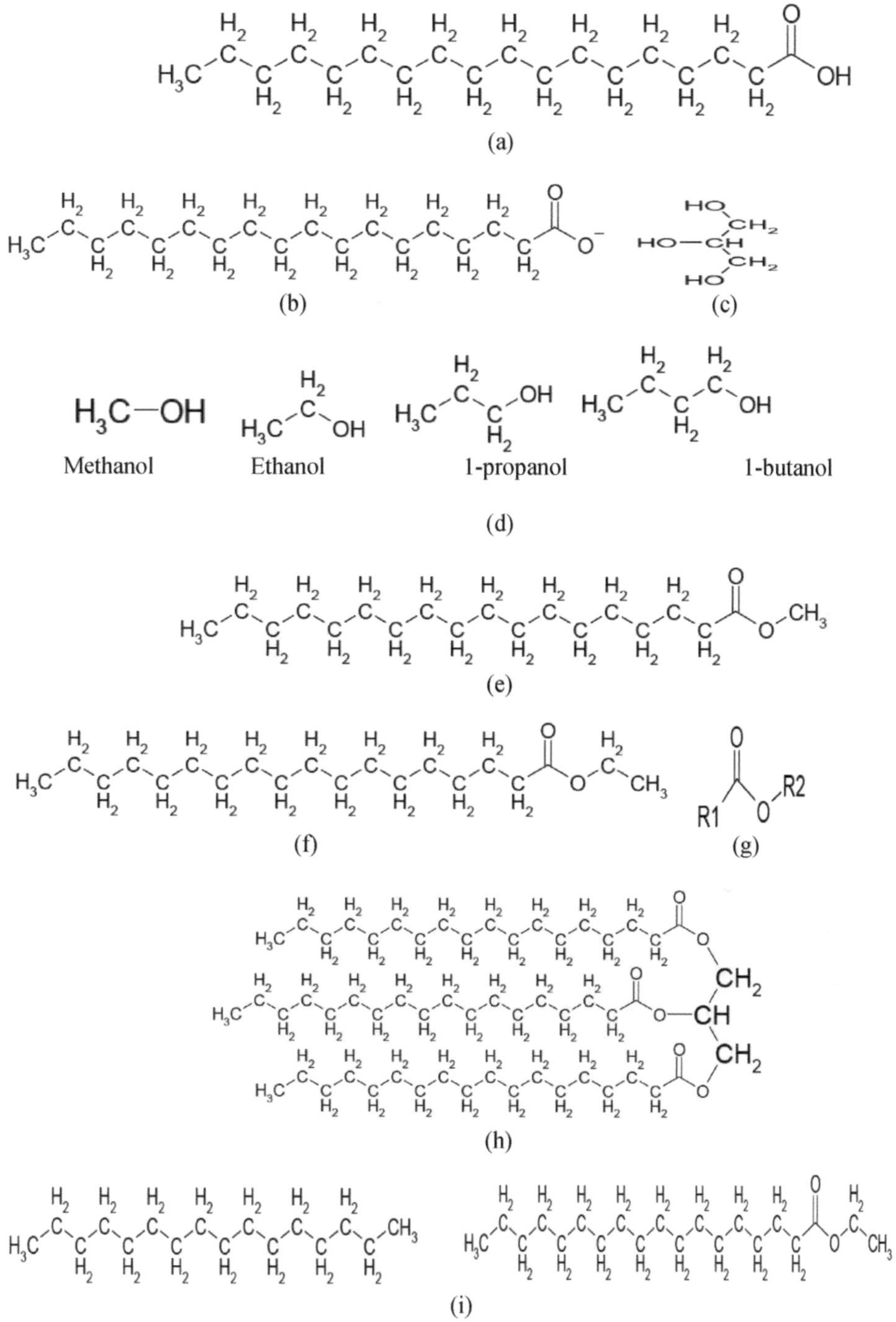

Figure 1.
Structures of chemical building blocks involved in transesterification process. (a) Idealized fatty acid; (b) idealized soap; (c) glycerol; (d) alcohols used in biodiesel production; (e) methyl ester; (f) ethyl ester; (g) generalized ester structure; (h) Trigyceride. (i) Cetane versus ethyl ester.

consecutive mechanisms using complex kinetic models. Such works on sunflower oil ethanolysis [6], jatropha oil methanolysis (Kuma et al., 2011), African pear seed oil [1], mixed crude oil palm oil methanolysis [7] buttress the above point. It is noteworthy that the complexity in kinetic models proposed in the above reports challenges their industrial translation while simplified kinetic models suffice for practical purposes. Consequently, methanolysis reaction has been proposed to constitute three consecutive irreversible stages, more especially by the usual condition of using high methanol to oil ratio (>3:1) which shifts the reaction methyl to the right [8, 9].

Terminalia catappa; belongs to *combietaccea* family with meridional Asian origin. It occurs in nature and widespread in the sub-tropical zones of India. It is called sea almond or tropical almond or Indian almond. In Nigeria, it is grown basically for

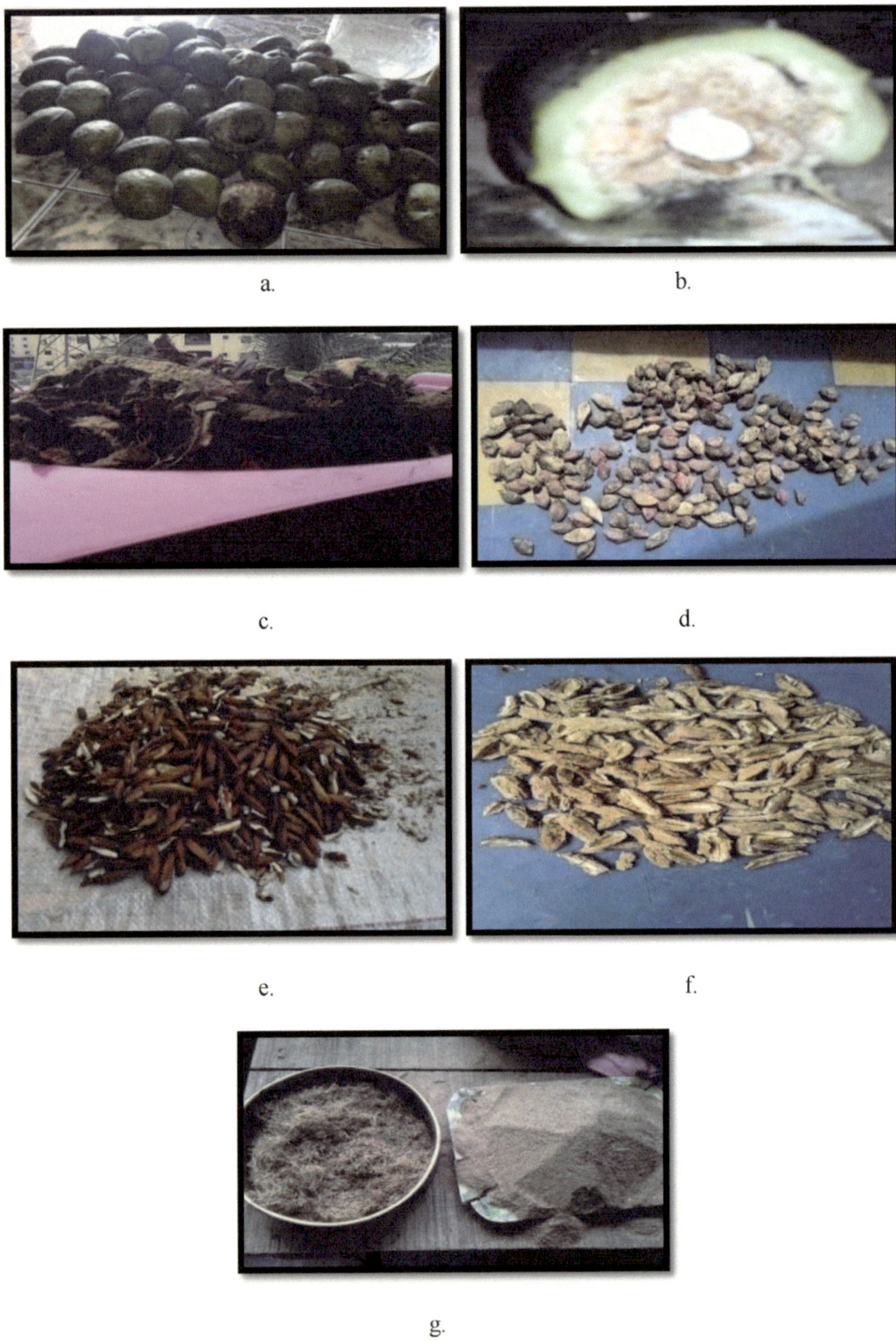

Figure 2.
Sea almond fruit biomass, a. the fruit, b. fruit cut section, c. dried fruit pulp, d. inner seed with coat. e. the seed, f. the fruit husk, g. the ground pulp (raffinnate and 600 μm particle size).

ornamental purposes [10]. It has been reported that the major works on *Terminalia catappa*, has focused mainly on the investigations of phyto-chemical, biological and medicinal application of its leaves, bark and fruit extracts with little or no attention to its seed oil industrial application [11]. Similar to other almonds like Iranian bitter almond and sweet almond, sea almond contains high amount of oil (>60%) [4, 12]. This is similar in quantity to what is observed in other established viable biodiesel feedstocks such as sunflower, peanut and rape seed [11]. Although empirical non-linear kinetics model of oil extraction as well as synthesis of transformer oil from seeds of *T. catappa* has been reported [11], process optimization and the kinetics of its seed oil methanolysis based on irreversible model under consecutive mechanism has not been reported. A pictorial representation of the *T. catappa* is shown in **Figure 2**. It is therefore the aim of this study to investigate and establish the optimal conditions, chemical kinetics and thermodynamic data for the production of biodiesel from *T. catappa* (sea almond) for its biofuel application relevance. This research is believed to compliment *T. catappa* seed oil's bio-lubricant potential as previously reported [11]. Additionally, the relevant characterizations through the application of nuclear magnetic resonance, gas chromatographic - mass spectrometry, Fourier transform infrared spectrometry analysis of the biodiesel were conducted and reported.

2. Materials and methods

2.1 Reagents

All the reagents used were all of analytical grade and purchased from the popular BriDGe-Head Chemical market in Onitsha, Anambra State Nigeria.

2.2 Biomass collection and preparation

2.2.1 Sourcing of seeds/seed meal preparation

The ripped fruits were collected from Abakaliki city of Nigeria. They were subsequently washed to remove dirt before the pulp was peeled out to release the kernel. The kernels were placed on solar drier for one (1) week. The seeds were extracted by cracking the kernels. Electric milling machine was used to grind the seeds into micro-sized meals before being sieved using an electric powered mechanical sieve to obtain a fine size of the meal. The remaining moisture in the sieved ground meal was removed by further sun drying the meal for a period of 5 days.

2.3 Oil extraction and degumming

The oil extraction followed the same method previously applied by the authors [1] but with slight modification. The extracted oil was further degummed by mixing the raw oil with 3 wt% by weight of warm water and the mixture was mechanically agitator coupled with using magnetic stirrer for 30 minutes at a temperature of 60° C to ensure that the emulsifiers were easily separated from the oil [13].

2.4 Physico-chemical characterization of the oil

The quality of the seed oil was determined in accordance with Association of Official Analytical Chemist [14] method. Other properties such as moisture, viscosity and density content were ascertained by using oven method, Oswald viscometer apparatus and density bottle respectively. The ash content and the refractive index were also measured with Veisfar muffle furnace and Abbe refractometer respectively. All the analyses were repeated three times and the average values were calculated and reported.

2.5 Base methanolysis process

The process follows the approach previously applied in Ofoefule et al. [13] with slight deviations. The extracted and pre-treated oil (100 ml) was first preheated to 80°C for 30 min before adding sodium methoxide. Sodium methoxide is more effective than direct mixing of sodium hydroxide due to the fact that direct mixing of NaOH with methanol produces water through hydrolysis and this affects the biodiesel yield. Therefore, sodium methoxide was prepared using the method previously reported by the authors [1]. Then the seed oil mixed with sodium methoxide at methanol/oil molar ratio of 6:1 was kept at 65°C for 65 min. This process was conducted in a 500 ml reflux condenser fitted with heater and stirrer. The process was conducted at atmospheric pressure and 140 rpm.

The biodiesel mixed with glycerine was separated, washed and dried according to the method previously applied by the authors [1]. The percentage biodiesel yield was calculated by using Eq. (1)

$$FAME\ yield(\%) = \frac{W_{FAME}}{W_{seed\ oil}} x100 \tag{1}$$

where $W_{FAME=weight\ of\ fatty\ acid\ methyl\ ester\ after\ methanolysis}$

$W_{seedoil=weight\ of\ seed\ oil\ used\ for\ the\ base\ methanolysis}$.

2.5.1 Physico chemical characterization of the biodiesel

The necessary fuel related physico-chemical properties of the biodiesel produced were determined using ASTM and AOAC [14] standard methods. ASTM D standards were used to determine the kinematic viscosity, density, pour, cloud, flash points, acid value and calorific values while AOAC methods were used to determine specific gravity, Iodine value and refractive index. ASTM D-445 method, the density was determined by ASTM D − 1298 method. The pour, flash and cloud points determinations were done using ASTM D-97, ASTM D-93, ASTMD-2500b methods respectively while acid value was measured by ASTM D-664 method. The refractive index was determined using AOAC 921.08. The specific gravity was ascertained using AOAC 920.212 and iodine value using AOAC 920:159 while moisture content was obtained using air-oven method. The cetane index (CI), cetane number (CN) and higher heating values were ascertained using standard correlations previously applied in [13].

2.5.2 Chemical characterization of seed oil and biodiesel

2.5.2.1 Nuclear magnetic resonance (NMR) analysis

The ^{13}C NMR of the sample was recorded on a Bruker Am-400 spectrometer operating at 100.6 MHz. The gated decoupling pulse sequence was used with the following parameters: Number of seans 512, acquisition time 1.366 s pulse with 10.3 s delay time 1.0 s. FID (free induction decay) was transformed and zero filled to 300 k to give a digital resolution of 0.366 Hz/point. Proton nuclear magnetic resonance (^{1}H NMR) spectra were recorded by dissolving approximately 100 mg sample 1 ml of deuterated chloroform solution and analysis using a Brucker model AC-250 spectrometer. Chemical shifts were measured in ppm downfield from internal tetramethyl siltane. The following instrumental parameters were applied. Spectrum width – 5000 Hz; acquisition time – 3.2775; delay time – 1 s and pulse width – 7 μsec.

2.5.2.2 Fourier transform infrared spectroscopic analysis of the oil and biodiesel

FT-IR analysis was performed to monitor the functional groups in the seed oil. The mid infrared spectra of oil samples were obtained in Fourier transform spectrometer by IR Affinity-1 Shimadzu, model No: 3116465. The FT-IR has SN ratio of its class of 30,000:1, 1 minute accumulator in the neighborhood of 2100 cm^{-1} peak to peak with a maximum resolution of 0.5 cm^{-1} in the region of 400 cm^{-1}-4000 cm^{-1}. It has microlab software as supporting software. The method of sample introduction was through sample cell. Cleaning of the cell was done with trisolvent mixture of acetone-toluene-methanol before background collection. About 0.5 ml of the sample (oil) was taken using the sample cell and introduced into the cell unit of the system. The scan results were obtained on the incorporated computer system as spectra. The peaks of the spectra obtained were identified and interpreted to identify the functional groups in the molecules of the oil with the aid of structure correlation chart [15].

2.5.2.3 Gas chromatographic-mass spectroscopic (GC: MS) analysis of the fatty acid profile of the biodiesel

The process followed the method reported by Esonye et al. [4]. The fatty acid composition of the biodiesel samples was in accordance with AOAC official method Ce2–66 using GCMS-QP2010 plus, Shimadzu. GC–MS is faster than the conventional GC; it equally provides molecular weight information and requires an aliquot sample. The GC–MS fragments the analyte to be identified on the basis of its mass and the column was calibrated by introducing methyl ester standards while good separations were achieved by diluting the sample in a little quantity of ethyl acetate. In this study, hydrogen served as the carrier gas and its flowrate was controlled at 41.27 ml/min while the flowrate of the column was 1.82 ml/min. Oven temperature was fixed at 80°C prior ramping up at 6°C/min and then up till 340°C. The Peaks identification was carried out by comparing their retention time and mass spectra with Mass Spectra Library (MSL) [16].

2.6 Optimization using RSM-desirability function techniques

2.6.1 Design of Experimental and Statistical Analysis

Central composite design (CCD) was applied in developing the design of experimental (DOE) for the base methanolysis of the *Terminalia catappa* seed oil. The matrix of the DOE based on the full factorial pattern provided sixteen (16) factorial points, eight (8) axial points and six (6) center points and these clearly present the required information on the inner conditions of the experimental circle. Design expert 7.0.0 software was employed for the design of the four (4) independent variables (n = 4), each with two (2) different levels. The total number of experiments (N) was worked out as $N = (n^2 + 2n + n_c) = 16 + 2(4) + 6 = 30$. This includes the standard 2n factorial points with their origin at the centre, 2n axial points fixed at a distance α from the centre to generate the quadratic terms and n_c replicate points at the centre. After defining the range of each of the process variable, they were coded to lie at ±1 for the fractional points, 0 for the centre point, ±α for the axial points. The numerical values of the variables were transferred into their respective coded values as shown in Eq. (2). The factor levels were coded as -α to +α as shown in the **Table 1** based on fuel factorial composite design (FFCDD). X_{min} (−α) and X_{max} (+α) are minimum and maximum values of X respectively, −1 and + 1 have a level of variance of $(X_{min} + X_{max})/2$ $(X_{max} - X_{min})/2b$ and 0 has a level of variance of $(X_{min} + X_{max})/2$. The effects of selected factors on the biodiesel yield were investigated based on the experimental conditions of the thirty set that were conducted. The main operating conditions (reaction time, alcohol to oil molar ratio, catalyst weight and reaction temperature) that conventionally affect methanolysis for biodiesel production were studied. **Table 1** contains the levels and range of the four independent variables. The variables range was chosen based on results obtained from previous works [17]. The presence of a clear curvature for the methanolysis resulted in selecting a second-order (Eq. (3)) for the transesterification [13].

$$X_i = \frac{2X - (X_{max} - X_{min})}{X_{max} - X_{min}} \tag{2}$$

$$Y = \beta_0 + \sum_{i=1}^{n} \beta_i x_i + \sum_{i=1}^{n} \beta_{ii} x_i^2 + \sum \sum_{i=1}^{n} \beta_{ij} x_{ij} \tag{3}$$

where, Xi - required coded value of a variable, X_{min} and X_{max} - the low and high values of X respectively, Where β_0 - a constant, β_i - the linear coefficient, β_{ii} – the quadratic coefficient, β_{ij}-interactive coefficients, X_i and X_{ij} are the uncoded independent variables and Y- predicted response (%). The fitted quadratic model equations obtained from regression analysis were used for the successful development of the response surface plots. The desirability function method was employed in order establish an efficient approach for achieving maximum FAME production. The application of one side transformation (Eq. (4)) followed by overall desirability (D) (Eq. (5)) using univariate technique was adopted [5, 13].

$$d_i = \begin{cases} 0 \; Y_i \leq Y_{i-min} \\ \left[\dfrac{Y_i - Y_{i-min}}{Y_{i-max} - Y_{i-min}}\right] Y_{i-min} < Y_i < Y_{i-max} \\ 0 \; Y_i \geq Y_{i-max} \end{cases} \tag{4}$$

$$D = \left(d_1^{w1} d_2^{w2} d_3^{w3} d_4^{w4} d_5^{wn}\right)^{1/\sum wi} \tag{5}$$

Parameters/Units	Symbols	Coded levels				
		-α	-1	0	1	+α
Temperature (°C)	X_1	30	40	50	60	70
Catalyst conc. (%wt)	X_2	0.5	1.0	1.5	2.0	2.5
Reaction time (min.)	X_3	45	50	55	60	65
Alcohol/Oil molar ratio	X_4	3:1	4:1	5:1	6:1	7:1

Table 1.
Variables, their symbols and CCD coded levels for Terminalia catappa *seed oil methanolysis.*

Where d_i is individual response desirability, Y_i is the response values, $Y_{i\text{-}min}$ is the minimum acceptable value for response i and $Y_{i\text{-}max}$ is the maximum acceptable value for response i. D is the overall desirability, wi is a weighed composite desirability.

The statistical methods used to ascertain the degree at which the models represent the experimental data were done by determining the coefficient of determination, (R^2) adjusted coefficient of determination (Adj. R^2), the mean squared error (MSE), root mean squared error (RMSE), the standard error prediction (SEP) and average absolute deviation (AAD) [13].

2.7 Chemical kinetic study

The rate of reaction and its mechanism as regards to the methanolysis process of the seed oil were investigated by considering irreversible conditions.

2.7.1 Equation of methanolysis reaction

It has been reported that the conventional transesterification mechanism could be represented by three consecutive irreversible [8] reactions as represented in Eqs. (6)–(8) with Eq. (9) being the summary of the Equations.

$$TG + AOH \underset{\rightarrow}{k_1} DG + BD \quad (6)$$

$$DG + AOH \underset{\rightarrow}{k_2} MG + BD \quad (7)$$

$$MG + AOH \underset{\rightarrow}{k_3} Gl + BD \quad (8)$$

$$\begin{array}{ccccccc} CH_2OCOR & & & & CH_2OH & & R'COOR \\ CHOCOR & + & 3ROH & \leftrightarrow & CHOH & + & R''COOR \\ CH_2OCOR & & & & CH_2OH & & R'''COOR \\ \text{Triglycerides} & & \text{Alcohol} & & \text{Glycerol} & & \text{Biodiesel} \end{array} \quad (9)$$

Where MG is monoglycerides, DG is Diglyceride, TG is Triglyceride, Gl is Glycerol, AOH is alcohol and BD is Biodiesel.

2.7.2 Irreversible model assumptions

Since simplified kinetic models suffice for practical purposes, experimental data were processed under the following assumptions [2, 8, 9]:

1. The methanolysis reaction is constituted by three consecutive stages but assumed irreversible because of the excessive presence of methanol in the reaction [9].

2. The free fatty acid neutralization was insignificant since the free fatty acid was negligible.

3. The saponification reaction was considered insignificant because of low acid value of the oil.

2.7.3 The kinetic experimental conditions

Kinetics experimental design (KED) of the methanolysis process of the sea almond seed oil followed the method previously reported by the authors in

Esonye et al. [1] with slight deviations to ascertain the kinetics and thermodynamic requirements. To examine the temperature dependency of the reaction rate constants, three (3) level temperatures (55–65°C) and twelve (12) intervals of reaction time (0-100 min) were considered at 6:1 alcohol (methanol)/sea almond seed oil molar ratio. About 2 ml aliquot sample were withdrawn at specified time intervals from the reactor, introduced into a test tube in an ice bath to quench the reaction. The content of the composite sample was obtained using a gas chromatography [1]. The G.C was equipped with split/splitless injection system operating at 185 degree Celsius, split ratio of 100:1, sample volume of 0.3 μL. High purity hydrogen gas was used as drag.

2.7.4 Second: order irreversible kinetic model

The best kinetic model for an irreversible model has been proposed to be a second-order based on TG hydrolysis especially during the early stages of the reaction [8]. To test the above report, a model developed based on TG hydrolysis and the second-order reaction rate for TG would be as shown in Eq. (10) [18].

$$\frac{-d[TG]}{dt} = k[TG]^2 \tag{10}$$

Resolving Eq. (10) further yields Eq. (11).

$$k_{TG}t = \frac{1}{[TG]} - \frac{1}{[TG0]} \tag{11}$$

Where k is the overall rate constant, t is the reaction time, TG_0 is the initial triglyceride concentration.

A plot of reaction time (t) against $\frac{1}{[TG]}$ will give a straight line if the model is valid. Where k is the overall rate constant, t is the reaction time; TG_0 is the initial triglyceride concentration. A plot of reaction time (t) against $\frac{1}{[TG]}$ will give a straight line if the model is valid. Similar approach was applied on the monoglycerides and diglycerides hydrolysis to get Eqs. (12) and (13).

$$k_{DG}t = \frac{1}{[DG]} - \frac{1}{[DG_0]} \tag{12}$$

$$k_{MG}t = \frac{1}{[MG]} - \frac{1}{[MG_0]} \tag{13}$$

2.7.5 First-order irreversible kinetic model

To determine the kinetics of the reaction based, the effect of reaction temperature and time were measured. It was assumed that the catalyst was used in sufficient amount with respect to oil to shift the reaction equilibrium towards the formation of fatty acid methyl esters. Thus, the reverse reaction could be ignored and change in concentration of the catalyst during the course of reaction can be assumed to be negligible [19]. Also, since the concentrations of both DG and MG were found to be very low (DG < 2.9 wt%, MG < 1.45 wt%) compared to those of TG (TG > 94 wt%) in the crude vegetable oils used in this research, the reaction could be assumed to be a single-step transesterification [20]. Therefore, the rate law of the transesterification reaction for forward reaction can be expressed by Eq. (14).

$$-r_{TG} = \frac{-d[TG]}{dt} = k'.[TG].[ROH]^3 \tag{14}$$

Where [TG] is the concentration of triglycerides and [ROH] that of methanol and k′ is the equilibrium rate constant. This overall reaction follows a second-order reaction rate law. However, due to the high molar ratio of methanol to oil, the change in methanol concentration can be considered as constant during reaction. This means that by taking methanol in excess, its concentration does not change the reaction order and it behaves as a first-order chemical reaction. Hence, the reaction would obey pseudo-first order kinetics [19] and finally, the rate expression can be written as in Eq. (15).

$$-r_{TG} = \frac{-d[TG]}{dt} = k.[TG] \tag{15}$$

Where k is modified rate constant and k = k′$[ROH]^3$. Assuming that the initial triglyceride concentration was $[TG_0]$ at time t = 0, and at time t it falls down $[TG_t]$. The integration of Eq. (15) for t = 0, [TG] = $[TG_0]$ and at t = t, [TG] = $[TG_t]$ gives Eq. (16):

$$-\ln [TG] + \ln [TG_0] = kt \tag{16}$$

In order to test the rate equation in Eq. (16), the experimental data were fitted to a straight line while the coefficient of determination was ascertained. A plot of –ln [TG] against time was obtained.

2.7.6 Thermodynamic requirement

In order to ascertain the process thermodynamic requirement, the values of rate constants were used to determine the Arrhenius activation energy from the plots of reaction rate constant (k) versus the reciprocal of absolute temperature (T) (Eq. (17)). DG and MG relationship with time followed the same trend with that of TG.

$$\log k = k_0 - \frac{E_a}{2.303R}\left(\frac{1}{T}\right) \tag{17}$$

Where E_a = Activation energy, R = Gas constant (8.314×10^{-3} J/Kmol), K = rate constant, K_O = frequency factor.

3. Results and discussion

3.1 Physico-chemical characterization result

The fuel related properties of the biodiesel and its parent oil obtained from this work at the optimum conditions are presented in **Table 2**. The properties of the biodiesel compared well with the American standards, European specification and other feedstocks recently applied for biodiesel production [4, 21]. The viscosity of the sea almond compared well with standards and other similar varieties. This is very important for the efficiency of its engine application since many diesel engines used injection pumps that do not accept high viscous fluids that clog the fuel filteration units. Also, sea almond had a better cetane number than Iranian bitter

Parameters	Sea almond seed oil[1]	Sea almond seed oil FAME[1]	Sweet almond seed oil FAME[2]	Iranian bitter almond seed oil FAME[3]	Standards		
					ASTM D 9751	ASTMD 6751	EN 14214
Oil/Biodiesel yield (%)	60.57	94.21	94.90	—	—	—	—
Density (kg/m^3)	856.10	855.3	849.1	887	850	880	860–900
Moisture content (%)	0.66	0.02	0.02	—	—	—	—
Refractive index	1.4471	1.441	1.4402	—	—	—	—
Acid value (mgKOH/g)	2.701	0.37	0.46	0.44	0.062	0.50	0.50
Free fatty acid (%)	1.35	0.18	0.23	—	0.31	0.25	0.25
Iodine value (mgKOH/g)	38.11	27.11	28.02	117.29	42–46	—	120max.
Saponification value (mgKOH/g)	166.21	162.3	161.05	185.35	—	—	—
Ash content (%)	1.00	0.01	0.01	—	0.01	0.02	0.02
Kinematic viscosity (mm^2/s)	—	2.40	2.52	4.68	2.6	1–9-6.0	3.5–5.0
Smoke point (°C)	40	36	34	—	—	—	—
Fire point (°C)	—	40	40	—	—	—	—
Flash point (°C)	156	138	136	173	60–80	100–170	120
Cloud point (°C)	−3	-3	−2	10	−20	−3 to 12	—
Pour point (°C)	—	−7	−6	−3	−35	−15 to 16	—
Calorific value (KJ/Kg)	—	32,188.50	31,178.39	—	42–46	—	35
Conductivity (Us/CM)	—	0.45	0.40	—	—	—	—
Cetane index	—	72.0	73.0	—	—	—	—
Cetane number	—	70.60	70.40	44.6	40–55	47 min.	51 min.
Higher heating value (HHV)[a] (MJ/kg)	—	35.62	34.72	—	—	—	—
Higher heating value (HHV)[b] (MJ/kg)	—	41.66	40.76	—	—	—	—
Higher heating value (HHV)[c] (MJ/kg)	—	64.65	63.75	—	—	—	—

[a]*Based on flash point.*
[b]*Based on viscosity.*
[c]*Based on density, min-minimum, max- maximum.*
[1]*This study.*
[2]*[4].*
[3]*[21]*

Table 2.
Physico-chemical properties of the sea almond seed oil and its FAME, sweet almond biodiesel and Iranian bitter almond biodiesel versus standards.

almond but compared well with sweet almond variety and standard specifications. This shows that sea almond oil is less unsaturated than Iranian bitter almond sea oil which has been reported to have 84.7% unsaturation [21] against 55.32% from sea almond and 52.42% for sea almond. The iodine value of sea almond was observed to be five (5) times less than Iranian bitter almond. Although Iranian bitter almond biodiesel iodine value is similar to that of tiger nut oil, the low value of sea almond biodiesel iodine value indicates less unsaturation. It equally shows that sea almond biodiesel will be comparatively less prone to oxidation instability and glyceride polymerization that normally leads to formation of deposits. The flash point, cloud point and pour point of Iranian bitter almond were very high compared to standards and the values recorded for both sea and sweet almond varieties. It implies that Iranian bitter almond variety will be safer to transport and handle in terms of flammability status and as well as be less suitable for winter season operations when compared with the hazardous and cold flow properties of sea almond. The parent oil characteristics of sea almond exhibited improved properties as a result of the base methanolysis [1].

3.2 FTIR characterization result

Table 3 contains peaks identified from the spectrum of the sea almond seed oil and its biodiesel. The band regions between 1734.60 cm^{-1}-1860.18 cm^{-1} and 1734.60 cm^{-1} - 1819.44 cm^{-1} for the oil and its biodiesel respectively can be ascribed to the stretching vibrations of C=O group. It shows the conversion of the triglyceride in the parent oil to biodiesel (methyl esters). Also, the specific bands of 2421.18 cm^{-1} and 2411.21 cm^{-1} appear with alkenes group for triglyceride and its biodiesel respectively. Also, the band regions between 3373.44–3495.22 cm^{-1} and 3365.18–3598.44 cm^{-1} for the parent seed oil and its biodiesel respectively can be ascribed to single-bonded hydroxyl group (O–H) stretching vibrations, appearing at high energy positions [4]. The single bond functional group O-H was observed to be prevalent in the biodiesel with stretch vibrations [4]. The presence of water molecule was evidenced by the hydrogen bonding [22]. The presence of C-H at 1357.64, 1474.28 and 1522.72 cm^{-1} regions of the biodiesel spectrum can be attributed to the properties such as pour and cloud points that influence the performance of biodiesel during cold weather engine operation [22]. However, the presence of carbon to

Sea almond seed oil			Sea almond seed oil biodiesel		
Wave number (cm^{-1})	**Type of Vibration**	**Functional group**	**Wave number (cm^{-1})**	**Type of vibration**	**Functional group**
892.50	Bending	=C-H	892.50	Bending	=C-H (alkenes)
1076.70	Bending	C-O-C	1041.96	Stretching	C-O
1188.64	Stretching	C-O	1134.60	Split rocking	C-O
1317.66	Bending/rocking	CH_2	1197.20	Split rocking	C-O
1474.28	Bending/rocking	CH_2	1317.66	Bending/Rocking	CH_2
1500.50	Bending/rocking	CH_2	1474.28	Bending/Rocking	CH_2
1734.60	Stretching	C=O	1555.12	Bending/Rocking	CH_2
1860.18	Stretching	C=O	1734.60	Stretching	C=O
2421.18	Symmetrical/ Stretching	C=C	1819.44	Stretching	C=O
3373.44	Stretching	O-H	2411.21	Symmetrical/ Stretching	C=C
3495.74	Stretching	O-H	3365.18	Stretching	O-H
			3598.44	Stretching	O-H

Table 3.
FT-IR main characteristic band positions for se almond seed oil and its biodiesel.

carbon (C=C) unsaturated bonds can cause the biodiesel samples to remain in liquid state but may be liable to poor storage stability due to oxidation. This implies that the biodiesel would not need cold flow improver for better performance. All the absorptions corresponding to C-O and C=O stretches indicate that the biodiesel product contains ester functional groups typical to any biodiesel type, while the following groups: C–H, C=H, and O–H indicated biodegradability of the oil and produced biodiesel [11]. Significant differences were effected by the ester groups. The specific peak that appeared at 892.50 cm^{-1} possesses bending type of vibrations appearing at low energy and frequency region in the spectra. It indicates the presence of = C–H functional groups [4]. It is part of fatty acid methyl ester with unsaturated bond in the seed oil and ester [23]. The specific peaks found in the region of 1088.80 cm^{-1} and 1197.20 cm^{-1} show split stretching and rocking vibrations of the carbonyl group (C–O) for the triglyceride and its methyl ester respectively [24]. The bending and rocking vibrations of methyl group in the parent oil and its methyl ester appeared between 1317.66–1500.50 cm^{-1} and 1317.66–1555.12 cm^{-1} respectively [25].

3.3 GC-MS characterization result

The various fatty acids present in the sea almond biodiesel are presented in **Table 4** in an increasing order of their retention time. A total of 38.14% saturated fatty acid, 39.92% monounsaturated fatty acid and 12.50% polyunsaturated fatty acids were found to be contained in the biodiesel The presence of high level of

Peak No.	Retention time (min.)	Fatty acid methyl ester	Amount (%)
1.	3.874	Capric acid	1.06
2.	4.017	Caprylic acid	1.14
3.	4.357	Stearic acid	1.24
4.	4.866	Eicosenic acid	8.14
5.	5.289	Erucic acid	0.75
6.	5.788	Palmitic acid	8.23
7.	6.729	Lignoceric acid	3.75
8.	7.243	Oleic acid	39.34
9.	8.922	α- Linolenic acid	9.07
10.	11.044	Palmtoleic acid	0.66
11.	11.281	Elaidic acid	1.09
12.	12.999	Arachidic acid	3.30
13.	14.569	Behenic acid	3.66
14.	16.888	Myristic acid	3.88
15.	18.367	Margaroleic acid	1.18
16.	22.223	Linoleic acid	0.72
17.	22.781	Gadoliec acid	0.12
18.	23.770	Lauric acid	1.66
19.	23.995	γ-linolenic acid	3.21
20.	23.875	Vaccenic acid	2.01

Table 4.
Fatty acid profile of the sea almond seed oil biodiesel.

monounsaturated fatty acids in methyl esters translates to high biodiesel quality [26]. Therefore, the high levels of monounsaturated fatty acids (39.92%) contained in the sea almond seed oil methyl ester is expected to make it possess excellent fuel qualities. Also, the higher the amount of unsaturated fatty acid in a biodiesel sample, the better the cloud point but lower the oxidation stability which implies that the higher composition of unsaturated fatty acids in the methyl ester (52.42%) would therefore enhance its cold flow properties [27]. It is reported that the high viscous nature of waste frying oils is because of their high saturated and less unsaturated fatty acids and this could cause micro-crystal formations that are dangerous to engine fuel injection units [28, 29]. Therefore, the application of biodiesel derived from the kernel seed of sea almond would possess no inherent viscosity problem. According to the present investigation, the cetane number of sea almond seed oil methyl ester is 63.39, and this shows the presence of high amount of monounsaturated fatty acids [30]. Methyl esters derived from animal sources has cloud point of about 17°C which is quite above 13°C obtained from palm oil sourced biodiesel while conversely feedstocks with lower concentrations of saturated fatty acids produces methyl esters with very low cloud point (< 0°C) [30]. Basically, biodiesel properties such as cetane number, kinematic viscosity, oxidative stability and cold flow properties are the specifications that are required to be satisfied and these have high relationship with the biodiesel fatty acid structural composition [31, 32]. Knothe [33] has reported that exhaust emission, and heat of combustion are likewise influenced by the fatty acid composition while methyl oleate is reported to be the most desired fatty acid to furnish produced biodiesel with the above expected fuel properties [1].

3.4 NMR characterization result

Nuclear magnetic resonance spectroscopy (NMR) is one of the instrumental analytical techniques used to quantify the conversion of triglycerides in vegetable oils into s [34, 35]. It is therefore, considered as one of the promising techniques for the characterization of biodiesel. The percentage conversion of the parent oil into its biodiesel using integration values for methoxy and α - carbonyl methylene protons [35] was found to be 95.7%%. Experimentally, the maximum sea almond yield obtained numerically as presented in **Table 2** and by GC maximum determination after 1 hr. were 94.21% and 93.01% respectively. All results are quite in good agreement and validate each other. The slight variation in conversion could be due to incomplete separation of FAME s from glycerine (by-product) and minor system errors as in the case of experimental and GC determinations respectively. The ^{1}H NMR spectrum of biodiesel from sea almond seed oil biodiesel is shown in **Figure 3a**. The specific peaks appearing at 0.452 ppm and 0.811 ppm for terminal methyl protons (C-CH_3) appears as singlet. From the ^{1}H NMR, the peak around 0.452 are from the terminal alkyl methyl in the s [36]. **Figure 3b** shows the ^{13}C NMR spectrum of biodiesel from the sea almond seed oil. The ^{13}C NMR shows significant aliphatic composition (CH_3) at the 24–28 ppm resonance [37] and for terminal carbon methylene at 17.774 pp. The peak at 124.629 ppm is typical of polycyclic aromatics structures [38]. Also, the peak at 167.288 ppm shows the presence of carbonyl carbon (-COO-) and O-aromatics (C-O) [34]. The peaks at 17.774–28.907 ppm could be attributed to terminal methyl groups. The unsaturation characteristics of s was confirmed by peaks appearing at δ124.629 ppm [34].

3.5 RSM optimization of sea almond seed oil methanolysis process

A central composite design (CCD) was applied to develop a correlation between the factors affecting transesterification reaction and the yield. The complete design

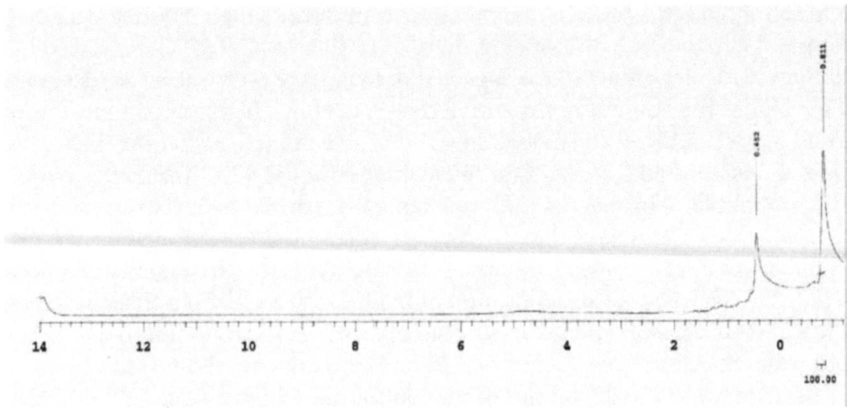

(a)

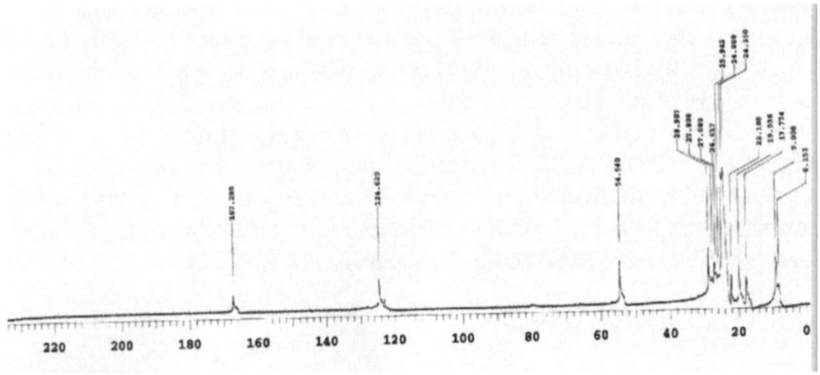

(b)

Figure 3.
(a) ^{1}H NMR spectrum of the biodiesel. (b) ^{13}C NMR spectrum of the biodiesel.

matrix, experimental and predicted responses is presented in **Table 5**. The experimental values of the content obtained were found to be in the range of ranged from 60 > actual value >95 wt %.

3.5.1 The RSM quadratic model ANOVA

The analysis of variance (ANOVA) of the RSM models (Linear, interactive linear, quadratic and cubic) were performed by considering the significance of the Fischer's F-value, lack of fit, degree of freedom (df) and R-squared (R^2). The result showed that the quadratic model best-satisfied the above set criteria. Other relevant appraisal methods involved the determination of coefficient of determination (R^2), adjusted coefficient of determination as well as coefficient of variation (C.V). These were applied to ascertain the adequacy of the model [13]. **Table 6** contains the effect of parameters using the second-order polynomial model. The following parameters X_1, X_2, X_3, X_1 X_2, X_1 X_3, X_2 X_4, X_1^2 and X_2^2 are found to be significant (**Table 7**). Since the parameters whose square are significant have more effect on the sea almond seed oil methanolysis [39], it implies that temperature, reaction time and catalyst had much effect on the studied response. The Model Fischer's F-value of 5.75 implies the model is significant and implies that there is only a 0.09% chance that a "Model Fischer's F-Value" this large could occur due to disturbance. The "Lack of Fit Fischer's F-value" of 0.2429 implies the Lack of Fit is not significant relative to the pure error. There is a 24.29% chance that a "Lack of Fit F-value" this

Run	Factor 1 X_1 (°C)	Factor 2 X_2 (wt %)	Factor 3 X_3 (min.)	Factor 4 X_4	Experimental value (%)	Predicted value (%)	Residual
1	40	1.0000	50	1:4000	79.8700	79.2754	0.5946
2	60	1.0000	50	1:4000	69.6400	71.6404	2.0014
3	40	2.0000	50	1:4000	65.9400	66.3137	−0.3737
4	60	2.0000	50	1:4000	68.7100	69.5587	−0.8487
5	40	1.0000	60	1:4000	83.9860	86.3597	−2.3737
6	60	1.0000	60	1:4000	86.7560	87.6047	−0.8487
7	40	2.0000	60	1:4000	83.0560	83.2380	−0.1820
8	60	2.0000	60	1:4000	85.8260	86.0330	−0.2070
9	40	1.0000	50	1:6000	66.8700	67.1823	−0.3123
10	60	1.0000	50	1:6000	70.6400	70.3273	0.3127
11	40	2.0000	50	1:6000	64.9400	65.9606	−1.0206
12	60	2.0000	50	1:6000	68.8100	69.7556	−0.9456
13	40	1.0000	60	1:6000	84.9860	84.2066	0.7794
14	60	1.0000	60	1:6000	85.7560	86.9016	−1.1456
15	40	2.0000	60	1:6000	85.0560	84.5349	0.5211
16	60	2.0000	60	1:6000	85.8260	87.3499	−1.5239
17	30	1.5000	55	1:5000	73.5780	76.3376	−2.7596
18	70	1.5000	55	1:5000	79.1180	79.3776	−0.2596
19	50	0.5000	55	1:5000	77.2780	79.7043	−2.4263
20	50	2.5000	55	1:5000	85.4180	86.0110	−0.4070
21	50	1.5000	45	1:5000	59.2320	61.6583	−2.4263
22	50	1.5000	65	1:5000	94.3640	94.1967	0.1673
23	50	1.5000	55	1:3000	76.3480	78.8357	−2.4877
24	50	1.5000	55	1:7000	76.0420	77.3425	−1.3005
25	50	1.5000	55	1:5000	75.9431	76.5521	−0.6090
26	50	1.5000	55	1:5000	75.9431	76.5527	−0.6090
27	50	1.5000	55	1:5000	75.9431	76.5521	−0.6090
28	50	1.5000	55	1:5000	75.9431	76.5521	−0.6090
29	50	1.5000	55	1:5000	75.9431	76.5521	−0.6090
30	50	1.5000	55	1:5000	75.9431	76.5521	−0.6090

Table 5.
The design matrix, experimental and predicted values of methanolysis process.

large could occur due to disturbance or noise. Non-significant lack of fit is good. It shows that the effect of most independent variables on the sea almond seed oil base methanolysis was significantly high. The non-significant lack of fit is good because it shows that the model will be well fitted [40]. The adequate precision compares the range of predicted values to the average prediction error. "Adeq Precision" measures the signal to noise ratio and a ratio greater than 4 is desirable (**Table 7**). The ratio of 8.148 obtained shows an adequate signal. The coefficient of variation is the ratio of the standard deviation of estimate to the mean value of the observed

Source	Sum of Squares	Df	Mean square	F value	p-value Prob > F	
Model	2157.2	14	154.09	5.75	0.0009	*Significant*
A- Temperature (X_1)	181.5	1	181.5	6.77	0.0200	*Significant*
B-Catalyst Conc. (X_2)	181.5	1	181.5	6.77	0.0200	*Significant*
C-Reaction Time (X_3)	190.2	1	190.2	7.09	0.0167	*Significant*
D-Metha/oil molar ratio (X_4)	28.17	1	28.17	1.05	0.3216	
AB- X_1 X_2	169	1	169	6.3	0.024	*Significant*
AC- X_1 X_3	144	1	144	5.37	0.035	*Significant*
AD- X_1 X_4	1	1	1	0.037	0.8495	
BC- X_2 X_3	36	1	36	1.34	0.2647	
BD X_2 X_4	196	1	196	7.31	0.0163	*Significant*
CD X_3 X_4	9	1	9	0.34	0.5709	
A^2 - X_1^2	1015.05	1	1015.05	37.86	< 0.0001	*Significant*
B^2 -X_2^2	304.76	1	304.76	11.37	0.0042	*Significant*
C^2 - X_3^2	92.19	1	92.19	3.44	0.0835	
D^2 - X_4^2	48.76	1	48.76	1.82	0.1975	
Residual	402.17	15	26.81			
Lack of Fit	319.33	10	31.93	1.93	0.2429	not significant
Pure Error	82.83	5	16.57			
Cor Total	2559.37	29				

Table 6.
Sea almond seed oils FAME yield response surface quadratic model ANOVA.

Std. Dev.	5.18	R^2	0.9429
Mean	76.77	Adj R^2	0.8562
C.V. %	6.75	PredR^2	0.6947
PRESS	958.64	Adeq Precision	8.148
RMSE	1.177	SEP	1.50150
MSE	1.217	AAD	0.5689

Table 7.
The regression model summary.

response and a measure of reproducibility and repeatability of the models [41]. Therefore, the C.V value of 6.75 shows the model is reasonably reproducible. Also, the R-squared of 0.9429 shows that more than 94% of the overall variability can be explained by the empirical models of the Equations. A given model significance can equally be validated when the standard deviation has a lower value than mean. Also, the smaller the PRESS-value the more the adequacy and significance of the model. Therefore, the PRESS-value obtained here supports the significance of the model. The adj. R-squared and the predicted R-squared values of 0.8562 and 0.6947 respectively for the quadratic model are in close agreement [42].

s/n	Operating variables	Sea almond[a]	Sweet almond[b]	Iranian Bitter almond[c]
1	Reaction time (min.)	58.52	65.00	60
2	Catalyst conc. (wt%)	2.04	1.5	1.4
3	Alcohol/oil molar ratio	4.66	5	9.7
4	Temperature (°C)	50.03	50	35
5	Predicted yield (wt%)	93.09	94.36	94.7
6	Experimental validated yield (wt%)	92.58	-	96.7

[a]*The present report.*
[b]*Esonye et al. [4].*
[c]*Mehdic and Kariminia [21].*

Table 8.
Optimized transesterification conditions for sea almond compared with sweet almond and Iranian bitter almond.

3.5.2 The RSM model equations

The chosen models based on coded, actual and significant terms are presented in Eqs. (18)–(20) respectively. The coded equation is useful for identifying the relative impact of the factors by comparing the factors coefficients, while the equation in terms of actual factors can be used to make predictions about the response for actual levels of each factor [40]. Analyzing the obtained model, it is observed that increase in the levels $X_1 X_2$, $X_1 X_3$, $X_1 X_4$ and $X_2 X_4$ results in a decrease in sea almond seed oil biodiesel yield [13].

$$\begin{aligned}\text{SASO FAME yield }(\%\text{w/w}) = {} & +86.83 + 2.75 * A + 2.75 * B + 0.75 * C + 1.08 * D \\ & - 3.25 * A * B - 3.00 * A * C - 0.25 * A * D \\ & + 1.50 * B * C - 3.50 * B * D + 0.75 * C * D \\ & - 6.08 * A^2 - 3.33 * B^2 - 1.83 * C^2 - 1.33 * D^2\end{aligned} \tag{18}$$

$$\begin{aligned}\text{SASO FAME yield }(\%\text{w/w}) = {} & -85.75000 + 2.75833 * \text{Temperature} \\ & + 21.37500 * \text{Cat Conc} + 2.42917 * \text{Reactionn Time} \\ & + 4.75000 * \text{Molar ratio} \\ & - 0.16250 * \text{Temperature} * \text{Cat Conc} \\ & - 0.015000 * \text{Temperature} * \text{Rxn Time} - 6.25000\text{E} \\ & - 003 * \text{Temperature} * \text{Molar ratio} \\ & + 0.15000 * \text{Cat Conc} * \text{Rxn Time} \\ & - 1.75000 * \text{Cat Conc} * \text{Molar ratio} \\ & + 0.037500 * \text{Rxn Time} * \text{Molar ratio} \\ & - 0.015208 * \text{Temperature}^2 - 3.33333 * \text{Cat Conc}^2 \\ & - 0.018333 * \text{Rxn Time}^2 - 0.33333 * \text{Molar ratio}^2\end{aligned} \tag{19}$$

$$\begin{aligned}\text{SASO FAME yield }(\%\text{w/w}) = {} & -85.75000 + 2.75833 * \text{Temperature} \\ & + 21.37500 * \text{Cat Conc} + 2.42917 * \text{Rxn Time} \\ & - 0.16250 * \text{Temperature} * \text{Cat Conc} \\ & - 0.015000 * \text{Temperature} * \text{Rxn Time} \\ & + 0.15000 * \text{Cat Conc} * \text{Rxn Time} \\ & - 1.75000 * \text{Cat Conc} * \text{Molar ratio} \\ & + 0.037500 * \text{Rxn Time} * \text{Molar ratio} \\ & - 0.015208 * \text{Temperature}^2 - 3.33333 * \text{Cat Conc}^2\end{aligned} \tag{20}$$

3.5.3 The production factors interactive effects

Figure 4A shows the 3D plot of interactive effects of reaction time and catalyst concentrations on sea almond biodiesel yield while keeping both the reaction temperature and methanol/oil molar ratio at constant zero (0) coded levels. The smoother curve of catalyst concentration axis on the 3D plots and its lesser quadratic coefficient p-values result clearly portrays that its quadratic is more significant than that of reaction time. It means that reaction time has less impact on the response than the catalyst amount. Optimum sea almond seed oil biodiesel yield was obtained at about of 58 minutes and 2.0 wt% catalyst amount and beyond these points the yield retarded. Similar range of reaction condition has been reported where highest yield of neem seed oil biodiesel was obtained at 60 min at all catalyst

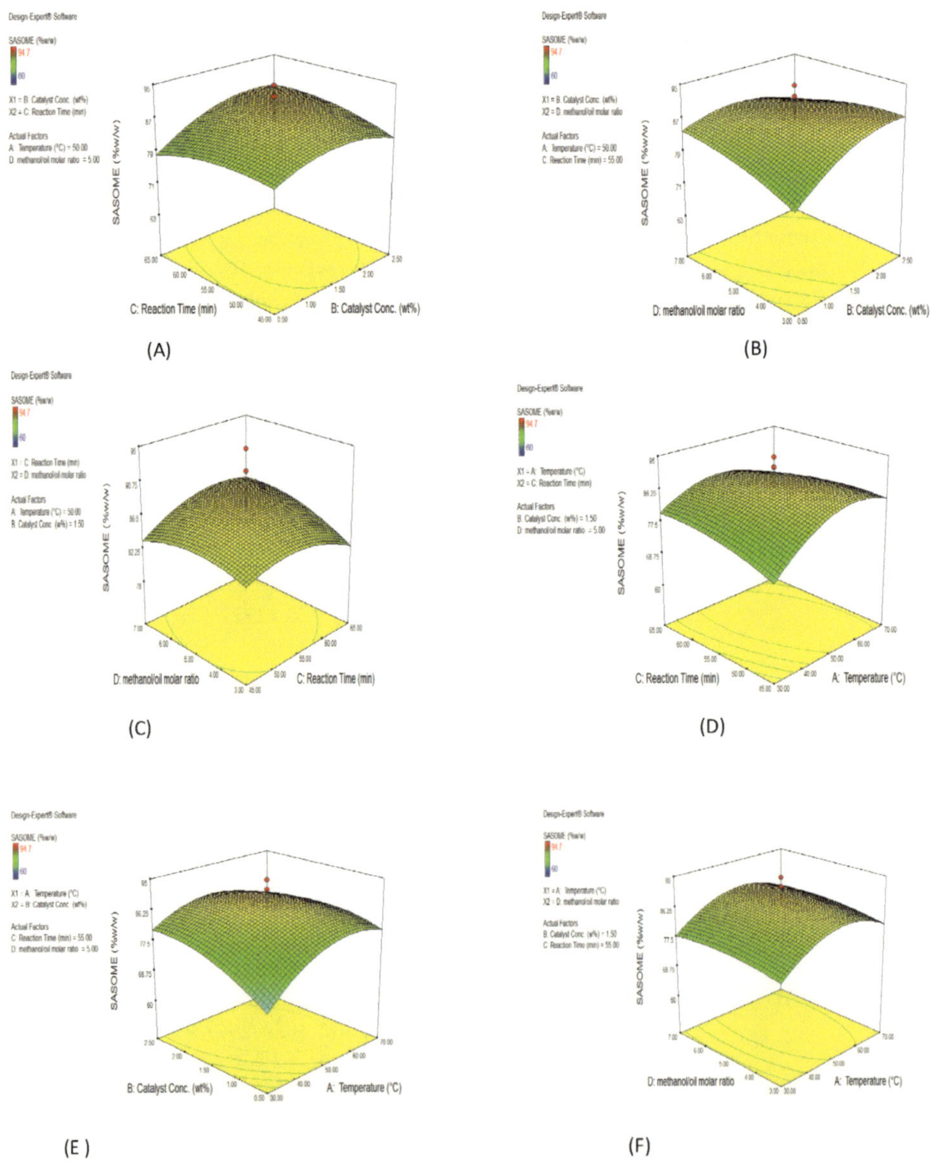

Figure 4.
The 3D response surface plot of the effects of the variables on sea almond FAME yield. (A). Reaction time and catalyst concentration. (B). Oil/methanol ratio and catalyst concentration. (C). Oil/methanol ratio and reaction time. (D). Reaction time and temperature. (E). Catalyst concentration and temperature. (F). Oil/methanol ratio and temperature.

concentration [39]. The reason could be because longer reaction time and excess catalyst promotes saponification reaction and increases in biodiesel viscosity respectively (Ofoefule, 2019). The impact of oil/methanol ratio and catalyst concentration while keeping other factors constant at 50°C and 55 minutes is represented in **Figure 4B**. The impact of both factors appears equal on the sea almond seed oil biodiesel yield and increase in both factors results in significant increase in the response. The response was observed to increase at all alcohol/oil molar ratios. However, below 2.5 wt% catalyst concentration showed increase effect on the response. Maximum yield was obtained at the highest catalyst concentration and molar ratio. Optimum yield was not attained by this combinations and this could be due to the fact that higher factors are required for them or the other factors kept constant at zero (0) levels requires shifts from the central points. Although literature reports that above these ranges, the yield decreased significantly even with increase in catalyst concentration, this could be that the excess catalyst (NaOH) reacts with methanol to form soap or produced emulsions that made the produced biodiesel had difficulty in the separation [43]. The feedstock studied here could have some deviating attributes or properties.

It was observed from **Figure 4C** that simultaneous increase in both oil/methanol molar ratio and reaction time resulted in yield increase until a certain point (6,1 and 60 min.) when it began to decrease. The smooth curves of both variables indicate that they had very significant effect on the yield of sea almond seed oil biodiesel. Both factors have almost the same impact on the biodiesel yield. Beyond these maximum points, increase in reaction time could have favored the backward reaction due to reduced concentration of the sea almond seed triglyceride while increase in molar ratio could have resulted in poor separation and recovery of glycerol [43]. This is because higher methanol content has been reported to promote high dissolution of the transesterification by-product which accelerates the reversible reaction [44]. From **Figure 4D**, the effect of reaction time and temperature while keeping other factors constant at 5.0 and 1.5 wt% for methanol/oil molar ratio and catalyst concentration respectively is presented. It shows that temperature has higher impact on the FAME yield than reaction time. The ANOVA results still show that the interactive term of temperature and reaction time was very significant while both the linear and quadratic terms of temperature were all more significant than those of reaction time similar to reports of Ofoefule et al. [13]. Basically, the higher

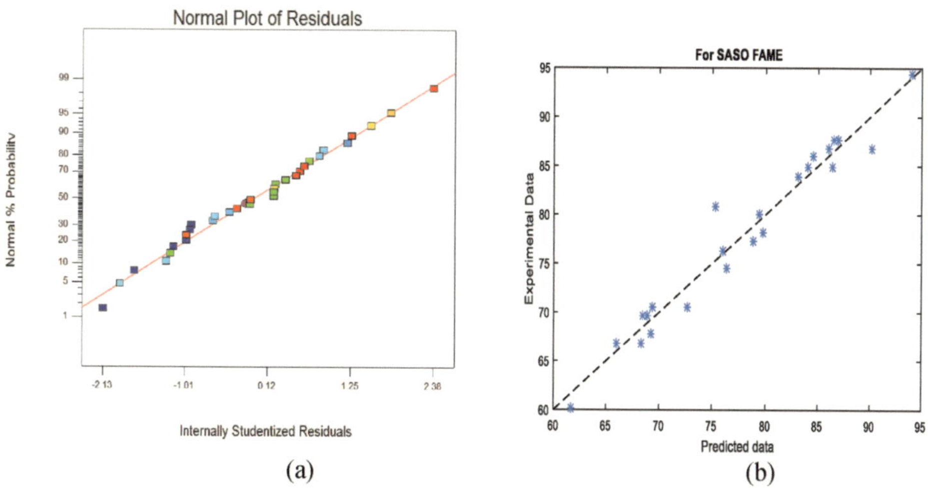

Figure 5.
(a) Normal probability plots of residuals and (b) linear correlation experimental and predicted values from sea almond seed oil methanolysis.

the temperature, the higher the reaction rate due to increase in the average kinetic energy of the reacting molecules according to Arrhenius theory [44]. The optimum temperature (60°C) would entail low cost of production as energy requirement for the seed oil methanolysis is comparatively low. Likewise, beyond 60 minutes reaction time, saponification might have been favored more due to less concentration of the reactants to push the reaction in the forward direction.

Figure 4E shows the 3D surface plot of the effects of catalyst concentration and temperature on the biodiesel yield of sea almond seed oil while keeping the reaction time and methanol/oil molar ratio constant. It shows the same trend with what was reported by Ofoefule et al. [13] on African pear seed oil biodiesel production optimization, although the catalyst concentration for the optimum yield in this report is 0.5 wt% less than what the previous report had presented. However, the explanation for the observed trend is due to increase in viscosity of the reaction composition at high catalyst concentration [13, 45]. **Figure 4F** shows the effects of oil/methanol molar ratio and temperature on the FAME yield. The catalyst concentration and reaction time was kept constant at 1.5 wt% and 55 minutes respectively. Temperature is found to have more significant impact on the response variable than methanol/oil molar ratio (as supported by the ANOVA result in **Table 6**). The FAME yield increased with increase in temperature irrespective of the value of the methanol/oil molar ratio. A reverse observation is possible if ethanol and different factor ranges were applied [43]. Optimum temperature was observed to be between 50 and 70°C in line with previous works [46].

The response values obtained by inserting the independent values are the predicted values of the model. These values are compared to the actual and experimental values. **Figure 5a** shows the normal probability plots of the residuals for clear investigations and diagnostic analysis. As it can be seen in **Figure 5b**, the data points were closely distributed along the diagonal axis. This implies that there is a good correlation between the actual and predicted values. This further corroborates the correlation between the R^2 and adjusted R^2 values. By implication, the CCD is well fitted into the model and has the capability of carrying out the optimization exercise for methanolysis of the seed oil.

The result of the optimized conditions for the optimum response of sea almond seed oil is presented in **Table 8** in comparison with the results previously reported by [4] and Mehdic and Kariminia [21] on sweet almond and Iranian bitter almond respectively. This was carried out using numerical optimization tool function of the Design Expert 7.0.0 version. The flexibility of the software enabled the generation of a total of 11 solutions together with their respective desirability. The selected best solution based on the best declared desirability of 1.00 represents the optimized process conditions where the sea almond seed oil FAME maximum response was obtained as 93.09 wt%. The chosen conditions were equally considered based on the economic point of view by taking into cognizance the impart of temperature on energy requirement, amount of catalyst and alcohol/oil molar ratio on the raw material cost and reaction time on the overall production cost. To confirm the model's adequacy, a replicate experiment was performed using the optimum points derived from the process variables and a validated yield of 92.58 wt% was obtained. The obtained result presents a good correlation between the predicted and actual biodiesel yield at the optimum levels. It is pertinent to compare optimized conditions with previous works in the literature. Here, the optimized *modus operandi* from *T. catappa* (sea almond) is compared with other reported biodiesel production processes on similar almond varieties: sweet almond and Iranian bitter almond. The conditions quite compared in yield, reaction time, and fairly on catalyst concentration. However, Iranian bitter almond biodiesel temperature of 35°C is found to be quite low compared with 50°C recorded for the other varieties. This could be due to the fact that its alcohol/oil molar ratio was about twice the values recorded for sweet almond and sea almond varieties and the catalyst applied for Iranian bitter was KOH against NaOH applied for the other varieties. Although, sweet almond had the highest reaction time, 7°C above sea almond and 5°C above Iranian bitter almond, sea almond from this study has about 0.5 wt% catalyst higher. Above all, the three almond varieties irrespective of their climatic origin and chemical composition have similar optimum conditions for the base methanolysis of their seed oils (**Table 9**).

3.6 Chemical kinetic study results

Figure 6ai-aiii shows the variation of the intermediates of the sea almond methanolysis with time. The result obtained by observing the trend is similar to that previously reported by the authors [1]. However, there is a difference between the maximum points of last intermediates. From this work, the values were 4.8 wt% at 1.0 min and 4.98 wt% at 2.0 min at 55°C, 4.65 wt% at 1.0 min and 4.82 wt % at 2.0 min at 60°C and 4.51 wt% at 1.0 min and 4.70 at 2.0 min at 65°C. The maximum points of the last intermediates (DG) previously reported on African pear seer oil were 4.59, 4.20 and 4.10 wt% at 55°C, 60°C and 65°C respectively [1]. This difference could be due to the difference in the parent oil chemical properties. However, the results compare in values. Also, **Figure 6b** shows that the effect of temperature on the FAME yield clearly follows an increasing trend. It was observed that the difference in the concentration of FAME, within the studied temperature ranges was not significant at respective reaction times. It implies that other factors other than temperature such as reaction time, mixing intensity, etc. had more effects on the seed oil TG conversion to s. This agrees with the result of the optimization where the ANOVA showed that reaction time was more significant than temperature.

3.6.1 Second order irreversible base transesterification model

Least-square approximation was applied, in fitting a straight line to the experimental data according to a model developed based on TG hydrolysis and the second-order reaction rate as shown in Eq. (21) ([8, 18]). In each case the coefficient of determination (R^2) was determined.

Glyceride	**Temperature (T)**		**k (wt%/min)**	**E_a (Kcal/mol.)**
	(°C)	**1/T x10^3(K^{-1})**		
TG→DG	55	3.05	0.00960 (R^2 = 0.98)	12.76
	60	3.00	0.01010 (R^2 = 0.99)	
	65	2.96	0.01610 (R^2 = 0.98)	
DG→MG	55	3.05	0.00838 (R^2 = 0.98)	15.83
	60	3.00	0.00845 (R^2 = 0.97)	
	65	2.96	0.01592 (R^2 = 0.97)	
MG→Gl	55	3.05	0.01650 (R^2 = 0.98)	22.43
	60	3.00	0.02930 (R^2 = 0.99)	
	65	2.96	0.04090 (R^2 = 0.98)	

Table 9.
Summary of the kinetics result for sea almond seed oil second-order irreversible methanolysis.

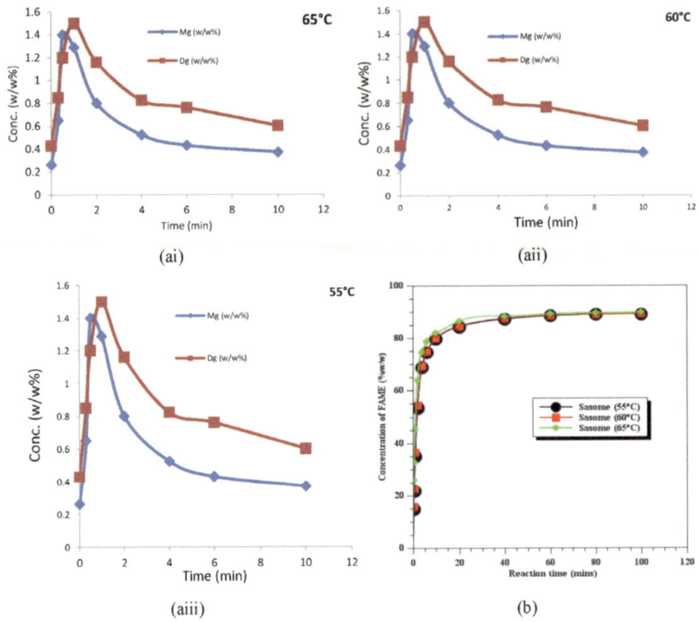

Figure 6.
(a) Progress of intermediates at various temperatures at the initial stage. (b) Effect of reaction temperature on the seed oil methanolysis.

$$\frac{-d[TG]}{dt} = k[TG]^2 \tag{21}$$

Integration of Eq. (21) gives Eq. (22).

$$k_{Tg}t = \frac{1}{[TG]} - \frac{1}{[TG0]} \tag{22}$$

Where k is the overall pseudo-rate constant, t is the reaction time, TG_0 is the initial triglyceride concentration.

A plot of reaction time (t) against $\frac{1}{[TG]}$ gave a straight line as shown in **Figure 7** with high values of coefficient (R^2) (**Table 9**) to show that the model is valid. The plot for the three temperatures (55, 60 and 65°C) is shown in **Figure 7a**, the slope is k_{TG} (wt%$^{-1}$min). It is observed that k fairly increased with temperature. Finally, activation energies of the reaction taking place were estimated using the calculated rate constants and temperatures at which they were observed in Arrhenius equation (Eq. (17)).

The DG and MG relationship with time followed the same trend (**Figure 7b** and **c**) with that of TG. There appears to be a very close similarity in the values of activation energy obtained in this study to the previous works [8] more especially in the Triglyceride and Diglycerides hydrolysis. However, the rate constants were found to be four (4) times higher and two (2) times lower than those reported by Darnoko and Cheryan [8] on palm oil base methanolysis and Reyero et al. [6] on sun flower base-ethanolysis. The choice of feedstocks, alcohol and other factors like temperature could have resulted in the slight differences. Also, the ratio constants increase with temperature follows a trend of $k_{TG} < k_{DG} < k_{MG}$ in values. After 60 mins reaction time, the highest conversion was above 90% and it is found to be in the same range with what many other researchers have reported [1]. The hydrolysis of TG to DG is observed to be the rate determining step since it is the slowest (with smallest k) while the DG conversion to glycerol is most favored by high temperature. It is observed that

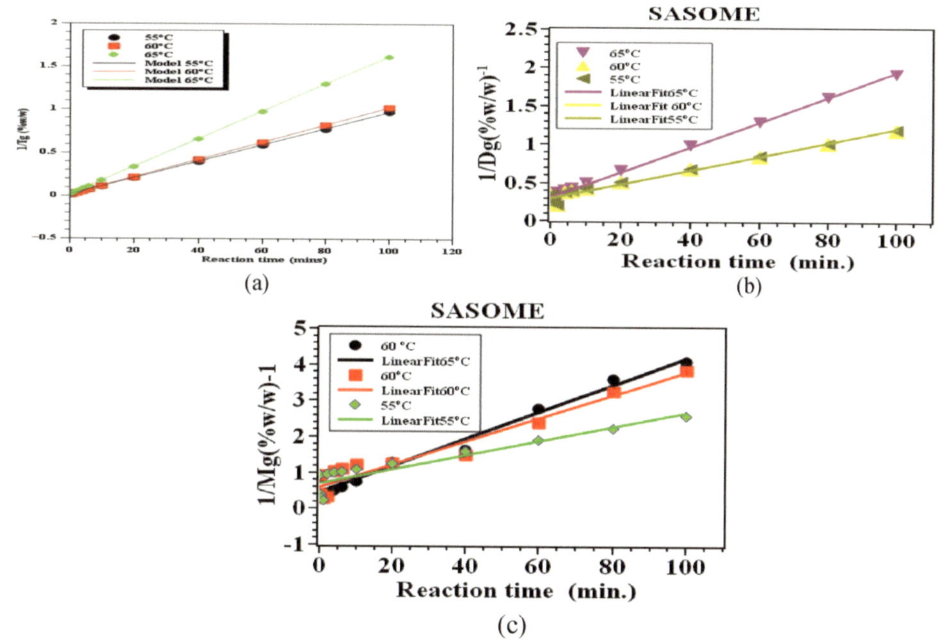

Figure 7.
Second-order reaction irreversible model of (a) triglycerides, (b) diglycerides and (c) monoglycerides hydrolysis.

Glyceride	Temperature (°C)	Reaction rate constant (min^{-1})	R^2
Triglyceride	55	0.0429	0.81
	60	0.0476	0.80
	65	0.0458	0.77

Table 10.
Summary of the rate constants for the first-order irreversible methanolysis.

all the steps have positive activation energy and this supports the endothermic characteristics of conventional transesterification process (**Figure 8**) [1].

3.6.2 First-order irreversible model

By ignoring the intermediate reactions of diglyceride and monoglyceride, the three steps have been combined in a single step [47]. However, due to the high molar ratio of methanol to oil, the change in methanol concentration can be considered as constant during reaction. This means that by taking methanol in excess, its concentration does not change the reaction order and it behaves as a first order chemical reaction [19]. The overall pseudo rate constants obtained from the slopes of the straight line plots of ln [TG] against t as shown in **Figure 9** are contained in **Table 10** for sea almond biodiesel. As can be seen from **Figure 9**, in the reactions conducted at 55, 60 and 65°C, there was a decrease in the coefficient of determination for the pseudo first-order kinetic model. **Figure 10** shows that the reaction at these temperatures does not fit the pseudo first-order reaction kinetic model better. This is supported by the lower values of coefficient of determination obtained from the first-order fitted plots ($R^2 < 0.80$) against high coefficient of determination obtained on the second-order irreversible kinetic model ($R^2 > 0.97$). Similar results have been reported on the kinetics of hydrolysis of *Nigella sativa* (*black cumin*) seed

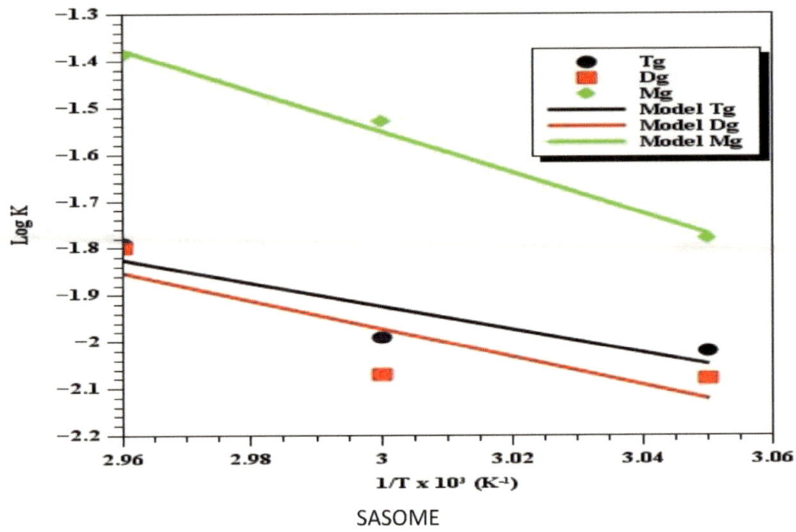

Figure 8.
Arrhenius plot of irreversible second order model reaction rate versus temperature.

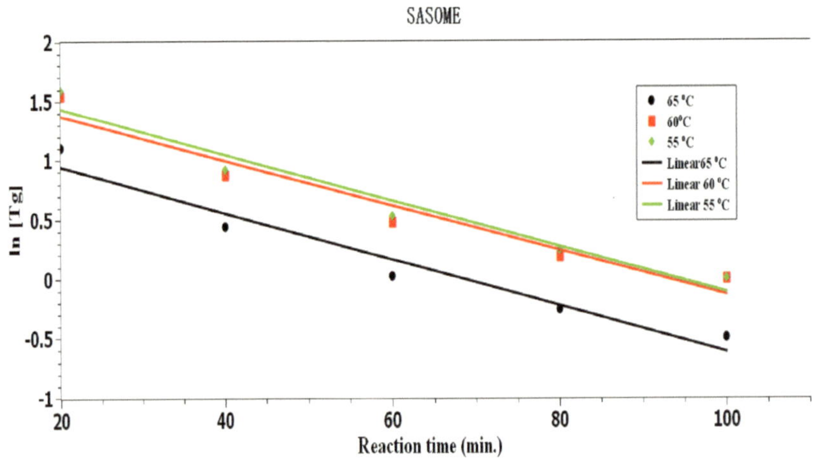

Figure 9.
First-order plot of the latter stage (from 20 minutes) triglycerides hydrolysis.

oil catalyzed by native lipase in ground seed where pseudo first-order rate equation at 20, 30 and 40°C; and the pseudo second-order equation at 50, 60 and 70°C [48]. Therefore, it could be that hydrolysis of some oils to s follows first-order irreversible kinetic models at low temperature ranges (20–40°C). The low temperature ranges is reported to favor the activity of native lipase better than at higher temperatures and this resulted in different mechanisms. But such low temperatures would not favor maximum ester yield in this study because they are far below the reported optimum temperature (Darnako and Cheryan, 2000). Darnako and Cheryan, 2000, has observed that at latter reaction stages (beyond 30 mins) of palm oil hydrolysis to, the first-order or zero-order reaction model is the best fitted. Similar observation was made on this study whereas from 20 minutes reaction, the reaction follows first-order model with high coefficient of determination ($R^2 > 0.94$). This is shown in **Figure 10**. These stages showed low reaction rate due to reduction in the reactants concentration. It implies that at low temperatures and latter stages of methanolysis of the vegeatble oils progesses very slowly and follow first-order kinetic model.

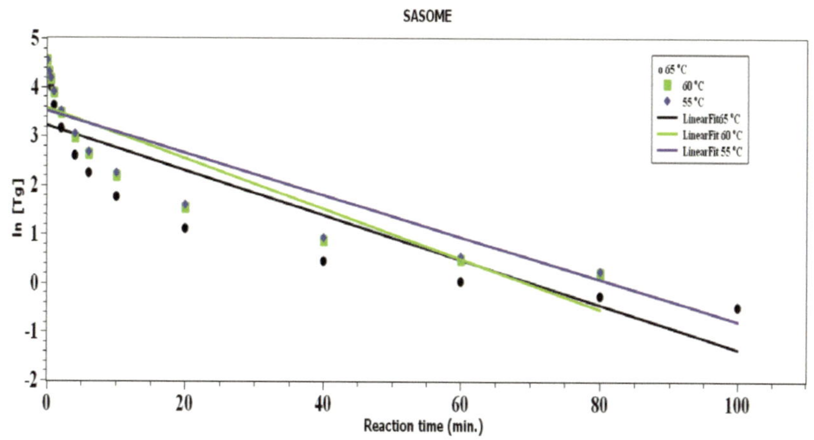

Figure 10.
First-order plot of the triglycerides hydrolysis.

4. Conclusion

The statistical optimization and chemical reaction kinetics of consecutive irreversible second order alkali- transesterification of *terminalia cattapa* seed oil has been successfully achieved and reported. RSM from Design Expert 7.0.0 version software was used for optimizing and predicting the process conditions in line with standard methodologies. The optimum conditions of base methanolysis process of the sea almond seed oil was obtained at favorable economic standpoint considering cheap materials requirement, low energy consumption and fast production rate. At low temperatures and latter stages, the methanolysis progresses very slowly and followed first order kinetic model but the irreversible second-order model of the power rate law best described the conversion of triglycerides with time at all stages. The data generated from the statistical optimization and chemical kinetics evaluations are found to be complimentary. The 's unsaturated characteristics would enhance its cold flow properties. The fuel properties of the biodiesel produced compared well with international standards. This research would help in commercial production of biodiesel from *T. cattapa* on industrial scale.

Acknowledgements

The authors would like to thank the staff and management of the PZ/NOTAP Chemical Engineering laboratory of Alex Ekwueme Federal University, Abakaliki, Nigeria for the availability of the laboratory facilities, apparatus and analytical equipment.

Conflict of interest

The authors hereby declare no competing financial interest.

Funding

This research did not receive any specific grant from any funding agent in public, commercial or not-for-profit-sectors.

Data availability

Research data are not shared

Notations

[Al]	Alcohol concentration
Al	Alcohol
DG	Diglycerides
[DG]	Diglycerides concentration
E_a	Activation energy (kcal/min)
FAAE	Fatty acid alkyl ester
FAEE	Fatty acid ethyl ester
FAME	Fatty acid methyl ester
Gl	Glycerol
[Gl]	Glycerol concentration
k_1-k_3	Rate constants (wt%/.min)
k_0	Frequency factor
MG	Monoglycerides
[MG]	Monoglycerides concentration
SASO	sea almond seed oil
SASOME	sea almond seed oil methyl ester
T	Temperature (K or °C)
TG	Triglyceride
[TG]	Triglyceride concentration

Author details

Chizoo Esonye[1]*, Okechukwu Donminic Onukwuli[2], Akuzuo Uwaoma Ofoefule[3], Cyril Sunday Ume[1] and Nkiruka Jacintha Ogbodo[1]

1 Department of Chemical Engineering, Alex Ekwueme Federal University, Ndufu-Alike, Abakaliki, Nigeria

2 Department of Chemical Engineering, Nnamdi Azikiwe University, Awka, Nigeria

3 Department of Industrial and Applied Chemistry, University of Nigeria, Nsukka, Nigeria

*Address all correspondence to: eso_vic@yahoo.com; esonye.chizoo@funai.edu.ng

References

[1] Esonye C, Onukwuli OD, Ofoefule AU. Chemical kinetics of a two-step Transesterification of Dyacrodes Edulis seed oil using acid-alkali catalyst. Chemical Engineering Research and Design. 2019a;**145**:245-257 http://dx.doi.org/10.1016/j.cherd.2019.03.010

[2] Patil P, Gude VG, Pinappu S, Deng PS. Transesterification kinetics of Camelina sativa oil on metal oxide cata; lysts under conventional and microwave heating condions. Chemical Engineering Journal. 2011;**168**:1296-1300

[3] Chatterjee S, Kumar A, Basu S, Dutta S. Application of response surface methodology for methylene blue dye removal from aqueous solution using low-cost adsorbent. Chemical Engineering Journal. 2012;**181-182**: 289-299

[4] Esonye C, Onukwuli OD, Ofoefule AU. Optimization of production from Prunus Amygdalus seed oil using response surface methodology and artificial neural networks. Renewable Energy. 2019b;**130**(62–71) http://dx.doi.org/10.1016/j.renene.2018.06.036

[5] Sarve A, Varma MN, Sonawane SS. Response surface optimization and artificial neural network modeling of biodiesel production from crude mahua (Mahua indica) oil under supercritical ethanol condition using CO_2 as co-solvent. Royal Society of Chemists Advances. 2015;**5**:69702-69713

[6] Reyero JG, Arzamendi S, Zabala ML, Gandia P. Kinetics of NaOH-catalyzed transesterification of sun flower oil with ethanol to produce biodiesel. Fuel Processing Technology. 2015;**129**: 147-155

[7] Jansri SS, Ratanawilu B, Allen ML, Prateepchaukul G. Kinetics of production from mixed crude oil by using acid-alkali catalyst. Fuel Processing Technology. 2011;**92**: 1543-1548

[8] Darnoko D, Cheryan M. Kinetics of palm oil transesterification in a batch reactor. Journal of the American Oil Chemists' Society. 2000;**77**:1263-1267

[9] Ilgen O. Reaction kinetics of dolomite catalyzed transesterification of conola oil and methanol. Fuel Processing Technology. 2012;**95**:62-66

[10] Weerawatanakorn M. Terminalia catappa seeds oil as a new dietary healthy oil source. In: The Annual Meeting of the International Society for Nutraceuticals and Functional Foods (ISNFF). Taipei, Taiwan; 2013

[11] Menkiti MC, Agu CM, Udeigwe TK. Kinetic and parametric studies for the extractive synthesis of oil from *Terminalia catappa* L. kernel. Reac Kinet Mech Cat. 2016. DOI: 10.1007/s11144-016-1101-y

[12] Monnet, Y. T, Gbogeuri A, Kouadio P, Koffi B, & Kouame, L. P. (2012) Chemical characterization of seeds and seed oil from mature Terminalia catappa fruits harvested in cote d'Ivoire. Int J Biosci10:110–124.

[13] Ofoefule AU, Esonye C, Onukwuli OD, Nwaeze E, Ume CS. Modeling and optimization of African pear seed oil esterification and transesterification using artificial neural network and response surface methodology comparative analysis. Industrial Crops and Products. 2019. DOI: 10.1016/j.indcrop.2019.111707

[14] AOAC (2000).Official methods of analysis, (13th edition), Association of Official Analytical Chemists, Washington, DC

[15] Furnish BS, Hannaford AJ, Smith PWG, Tatchell AR. Vogel's Textbook of

Practical Organic Chemistry. 5th ed. UK: Longman group; 1989. pp. 1412-1422

[16] Fu YJ, Zu YG, Wang LL, Zhang NJ, Liu W, Li SM, et al. Determination of fatty acid s in biodiesel produced from yellow horn oil by RP-LC-RID. Chromatographia. 2008;**67**:9-14

[17] Betiku E, Ajala SO. Modeling and optimization of Thevetia peruviana (yellow oleander) oil biodiesel synthesis via Musa paradisiacal (plantain) peels as heterogeneous base catalyst: A case of artificial neural network vs response surface methodology. Industrial Crops and Products. 2014;**53**:314-322

[18] Levenspiel O. Chemical Reaction Engineering. 3rd ed. McGraw-Hill; 1999

[19] Zhang L, Sheng B, Xin Z, Liu Q, Sun S. Kinetics of transesterification of palm oil and dimethyl carbonate for biodiesel production at the catalysis of heterogeneous base catalyst. Bioresource Technology. 2010;**101**: 8144-8150

[20] Kumar GR, Ravi R, Chadh A. Kinetic studies of base-catalyzed transesterification reaction of non edible oils to prepare biodiesel; the effect of co-solvent and temperature. Energy and Fuels. 2011;**25**:2826-2832

[21] Mehdic A, Kariminia H. Optimization of biodiesel production from Iranian bitter almond oil using statistical approach. Applied Energy. 2011;**88**(7):2377-2381

[22] Ndana M, Grace JJ, Baba FH, Mohammed UM. Fourier transform infrared spectrometyric analysis of functional groups in biodiesel produced from oil of recinum communis, hevea brasiliences and jatropha curcas seeds. Int. J. of Sc. Envi. & Tech. 2013;**2**(6): 1116-1121

[23] Saifuddin N, Refai H. Spectroscopy analysis of structural transesterification in biodiesel degradation. Research Journal of Applied Sciences, Engineering and Technology. 2014;**8**(9): 1149-1159

[24] Conceicao MM, Fermardes VJ, Araiyo AS, Farais MF, Santos IM, Souza AG. Thermal and oxidative degradation of castor oil biodiesel. Journal of Energy Fuel. 2007;**21**(1):1522-1527

[25] Gunstone ED. Rapeseed and canola oil: Production process, properties and uses. JAOCS. 2004;**33**:132-139

[26] Knothe G. Improving biodiesel fuel properties by modifying fatty ester composition. Energy and Environmental Science. 2009;**2**:759-766

[27] Sharma YC, Singh B. Development of biodiesel current scenarior. Review Sustain. Energy Rev. 2009;**13**(6–7): 1646-1651

[28] Esonye C, Onukwuli OD, Ofoefule AU, Ogah EO. Multi-input multi-output (MIMO) ANN and Nelder-Mead's simplex based modeling of engine performance and combustion emission characteristics of biodiesel-diesel blending CI diesel engine. Applied Thermal Engineering. 2019c;**151**:100-114 http://dx.doi.org/10.1016/j.applthermaleng.2019.01.101

[29] Knothe G. Determing the blend levels of mixtures of biodiesel with conventional diesel fuel by fibre-optic near infra red spectroscopy and H magnetic resonance spectroscopy. J. America Oil Chem. Sci. 2001;**78**:1025

[30] Koria L, Nithya G. Analysis of Datura stramonium Linn, biodiesel by GC –MS and influence of fatty acid composition on fuel related characteristics. Journal of Phytology. 2012;**4**(1):06-09

[31] Knothe G. Biodiesel and renewable diesel: A comparison. Progress in Energy

and Combustion Science. 2010;**36**(3): 364-373

[32] Mittlebach M, Remschnodt C. Biodiesel the Comprehensive Handbook. Vienna: Boersedruck Ges MBH; 2004

[33] Knothe G. Synthesis and characterization of long chain 1, 2-dioxo compounds. Chemistry and Physiscs of lipids. 2012;**50**:14-34

[34] Mushtaq A, Shazia S, Lee KT, Ahmed ZA, Haleema S, Muhammed Z, et al. Distaff thistle oil: A possible new non-edible feedstock for biogas energy. International Journal of Green Energy. 2014;***10***:1-27

[35] Naureen R, Tariq M, Yusoff I, Chowdhury AJK, Ashraf MA. Synthesis, spectroscopic and chromatographic studies of sunflower oil biodiesel using optimized base catalysed methanolysis. Saudi Journal of Biological Sciences. 2015;**22**:332-339

[36] Nagaraja YP, Venkatesh HS, Vinoyty MS. Production of biodiesel from poultry waste. International Journal of Pharma and Bio Sciences. 2014;**592**:791-799

[37] Novak JM, Lema T, Xing B, Gaaskin JW, Steiner C, Das KC, et al. Characterization of designer biochar produced at different temperatures and their effects on a loamy sand. Ani Environ Sc. 2009;**3**:195-206

[38] Novotry, E. H., deAzevedo E. R., Bonagamba J. J. Cunha, T. J. F, Mmadari B. E. de M. Benites, V. & Hayes M. H. B. (2007) Studies of the compositions of humic ascids from Amazoruan Dark Earth Soils. Environ. Sc. Technol 41: 400–405.

[39] Awolu OO, Layokun SK. Optimization of two-step transesterification production of biodiesel from neem (Azadirachta indica) oil. Int. J. of Energy and Environmental Engineering. 2013;**4**:39

[40] Ohale PE, Uzoh CF, Onukwuli OD. Optimal factor evaluation for the dissolution of alumina from azaraegbelu clay in acidic solution using RSM and ANN comparative analysis. South African Journal of Chemical Engineering. 2017;**21**:43-54

[41] Chen, G., Chen, J., Srinivasakannan, C., & Peng., J. (2011). Application of response surface methodology for optimisation of the synthetic nitrile from titania slag. App. Surf. Sci. 268 (7) 3068–3073,

[42] Uzoh CF, Onukwuli OD, Ozofor IH, Odera RS. Encapsulation of urea with alkyd resin-starch membranes for controlled N_2 release: Synthesis, characterization, morphology and optimum N_2 release. Process Safety and Environmental Protection. 2019;**121**: 133-142

[43] Silva GF, Fernando L, Camargo F, Andrea LO, Ferreira F. Application of response surface methodology for optimization of biodiesel production by transesterification of soybean oil with ethanol. Fuel Processing Technology. 2011;**92**:407-413

[44] Ayodele A, Ayoola KF, Hymore KF, Omonhinmin CA. Optimization of biodiesel production from selected waste oils using response surface methodology. Biotechnology. **2017**;**16**:1-9

[45] Tshizanga N, Aransiola EF, Oyekola O. Optimisation of biodiesel production from waste vegetable oil and eggshell ash. South African Journal of Chemical Engineering. 2017;**23**:145-156

[46] Odude VO, Adesina AJ, Oyetunde OO, Adeyemi OO, Ishola NB, Anietie Okon Etim AO, et al. Application of agricultural waste-based catalysts to Transesterification of esterified palm

kernel oil into biodiesel: A case of Banana fruit Peel versus cocoa pod husk. Waste and Biomass Valorization. 2017 https://doi.org/10.1007/s12649-017-0152-2

[47] Birla A, Singh B, Upadhyay S, Sharma Y. Kinetics studies of synthesis of biodiesel from waste frying oil using a heterogeneous catalyst derived from snail shell. Bioresource Technology. 2012;**106**:95-100

[48] Dandik L, Aksoy HA. The kinetics of hydrolysis of Nigella sativa (black cumin) seed oil catalyzed by native lipase in ground seed. JAOCS. 1992;**69** (12):1239-1240